Christin Röttiger

Altersvergleichende Darstellung equiner Oberkieferbackenzähne und angrenzender Strukturen mittels 3-Tesla-MRT und CT unter Berücksichtigung des Präparatzustandes

Cuvillier Verlag Göttingen
Internationaler wissenschaftlicher Fachverlag

Bibliografische Information der Deutschen Nationalbibliothek
Die Deutsche Nationalbibliothek verzeichnet diese Publikation in der Deutschen Nationalbibliografie; detaillierte bibliographische Daten sind im Internet über http://dnb.d-nb.de abrufbar.
1. Aufl. - Göttingen: Cuvillier, 2020
Zugl.: Hannover (TiHo), Univ., Diss., 2020

Nonnenstieg 8, 37075 Göttingen
Telefon: 0551-54724-0
Telefax: 0551-54724-21
www.cuvillier.de

1. Auflage, 2020
Gedruckt auf umweltfreundlichem, säurefreiem Papier aus nachhaltiger Forstwirtschaft.

ISBN 978-3-7369-7215-5
eISBN 978-3-7369-6215-6

Tierärztliche Hochschule Hannover

Altersvergleichende Darstellung equiner Oberkieferbackenzähne und angrenzender Strukturen mittels 3-Tesla-MRT und CT unter Berücksichtigung des Präparatzustandes

INAUGURAL – DISSERTATION
Zur Erlangung des Grades einer Doktorin
der Veterinärmedizin
- Doctor medicinae veterinariae -
(Dr. med. vet.)

vorgelegt von
Christin Röttiger (geb. Schoppe)
Lingen an der Ems

Hannover 2020

Wissenschaftliche Betreuung:

1. Prof. Dr. B. Ohnesorge
   Klinik für Pferde
   Tierärztliche Hochschule Hannover

2. PD Dr. A. Bienert-Zeit
   Klinik für Pferde
   Tierärztliche Hochschule Hannover

3. Dr. M. Hellige
   Klinik für Pferde
   Tierärztliche Hochschule Hannover

1. Gutachter: Prof. Dr. B. Ohnesorge

2. Gutachter: Prof. Dr. C. Pfarrer

Tag der mündlichen Prüfung: 05.05.2020

Meiner Familie

Ergebnisse dieser Dissertation wurden in international anerkannten Fachzeitschriften mit Gutachtersystem (peer review) veröffentlicht:

1. BMC Veterinary Research (am 06.09.2017):
   Comparison of computed tomography and high-field (3.0 T) magnetic resonance imaging of age-related variances in selected equine maxillary cheek teeth and adjacent tissues
   C. Schoppe, A. Bienert-Zeit, K. Rohn, M. Hellige, B. Ohnesorge

2. Acta Veterinaria Scandinavica (am 02.12.2019):
   Magnetic resonance imaging and computed tomography of equine cheek teeth and adjacent structures: comparative study of image quality in horses in vivo, post-mortem and frozen-thawed
   C. Röttiger, M. Hellige, B. Ohnesorge, A. Bienert-Zeit

Teilergebnisse der Dissertation wurden auf folgenden Fachkongressen präsentiert:

GEVA - GPM 2nd International Congress, 28.-29.10.2016, Berlin
Comparison of age-related variances in equine maxillary cheek teeth portrayed in CT and 3-T MR images
C. Schoppe, A. Bienert-Zeit, M. Hellige, B. Ohnesorge

15. IGFP-Kongress, 10.-11.03.2017, Niedernhausen
CT oder MRT? Bestmögliche Darstellung von Backenzähnen und anliegenden Strukturen bei Pferden unterschiedlicher Altersgruppen
C. Schoppe, A. Bienert-Zeit, M. Hellige, B. Ohnesorge

26th European Veterinary Dental Forum (EVDF), 18.-20.05.2017, Malaga (Spanien)
Age-dependent positional relations and possible ascending infections of selected equine cheek teeth portrayed by 3.0 Tesla magnetic resonance and computed tomographic imaging
C. Schoppe, A. Bienert-Zeit, M. Hellige, B. Ohnesorge

Inhaltsverzeichnis

Abkürzungsverzeichnis

3D	dreidimensional
bzw.	beziehungsweise
ca.	circa
CT	Computertomographie
et al.	et alii (und andere)
FOV	field of view (das untersuchte Bildfeld)
kV	Kilovolt
mAs	Milliamperesekunden
min	Minute
mm	Millimeter
MPR	multiplanare Rekonstruktion
MR	magnetic resonance
MRT	Magnetresonanztomographie/ Magnetresonanztomograph
N.	Nervus
n	Anzahl
PDL	Periodontal ligament (periodontales Ligament)
PD SPAIR	proton density weighted spectral attenuated inversion recovery
PDw	proton density weighted (protonengewichtet)
s.	siehe
SNR	signal-to-noise-ratio (Signal-zu-Rausch-Verhältnis)
STIR	short time inversion recovery
SPAIR	spectral attenuated inversion recovery
T	Tesla
T1w	T1 weighted (T1 gewichtet)
T2w	T2 weighted (T2 gewichtet)
WW	window width (Fensterweite)
WL	window level (Fensterlage)

1 Einleitung

Zahnerkrankungen und dentogen bedingte Nasennebenhöhlenerkrankungen sind in der Pferdemedizin aufgrund ihrer kostenintensiven Behandlung und der oftmals langanhaltenden Nachsorge von besonderer klinischer Relevanz. Die Ätiopathogenese apikaler Infektionen der Oberkieferbackenzähne ist bislang nicht in allen Fällen abschließend geklärt (BÜHLER et al. 2014). Die Diagnose basiert auf der klinischen Untersuchung des äußeren Kopfes, einer detaillierten orodentalen Befundung, einer Endoskopie der oberen Atemwege und weiterer bildgebender Untersuchungen. Klinische Anzeichen einer dentogen bedingten Sinusitis, wie einseitiger Nasenausfluss oder Schwellungen im Angesicht, treten in der Regel zeitlich deutlich versetzt zu den initialen pathologischen Veränderungen der Zahnstrukturen und des periodontalen Gewebes auf. Um frühzeitig Zahnerkrankungen erkennen und behandeln zu können, ist die bildgebende Diagnostik von endodontischen Erkrankungen (GERLACH et al. 2013), Infundibularkaries (VERAA et al. 2009) und apikalen Infektionen (CASEY et al. 2015) erheblich ausgeweitet worden.

Mit Hilfe röntgenologischer Untersuchungen lassen sich Befunde wie ein erweiterter Periodontalraum, eine Lyse der Zahnwurzel oder der Lamina dura, die Sklerose des Alveolarknochens oder auch die Füllung der Nasennebenhöhlen erheben (ROS 2011). Die Röntgenuntersuchung der Kopfregion ist in der Pferdemedizin etabliert und für die weiterführende Zahndiagnostik das bildgebende Verfahren der ersten Wahl (TREMAINE u. DIXON 2001, BARAKZAI 2011). Nachteilig ist jedoch, insbesondere im Bereich des Schädels, die Überlagerung komplexer anatomischer Strukturen, die die Interpretation der Befunde erschwert (PARENTE et al. 2011).

Die Computertomographie (CT) und die Magnetresonanztomographie (MRT) bieten den Vorteil, überlagerungsfreie Schnittbilder zu erstellen sowie die Möglichkeit der dreidimensionalen (3D) Darstellung. Mittels CT kann zudem eine multiplanare Rekonstruktion (MPR) angefertigt werden. Beide Bildgebungsmodalitäten sind nützliche Verfahren, um die unterschiedlichen Gewebe des Pferdekopfes abzubilden

(ARENCIBIA et al. 2000, BIENERT 2002, TUCKER u. SAMPSON 2007, GERLACH u. GERHARDS 2008).

Die Anwendung der CT zur Befundung dentogen bedingter Pathologien hat große Akzeptanz gefunden und sich in der Pferdezahnheilkunde als weiterführende Diagnostik etabliert (HENNINGER et al. 2003). Der Mehraufwand der CT ist gerechtfertigt, wenn sich keine zuverlässige röntgenologische Diagnose stellen lässt (DIXON 1997, BIENERT 2002). Der Nachteil der CT-Untersuchung besteht jedoch darin, dass Weichteilgewebe (z.B. Pulpen oder Ligamentum periodontale) im Vergleich zur Kernspintomographie nur eingeschränkt dargestellt werden können (GERLACH et al. 2013).

Die Möglichkeiten der MRT, eine exzellente Auflösung von Weichteilstrukturen zu bieten, werden in der Humanmedizin bereits zur Beurteilung von apikalen Infektionen (GEIBEL et al. 2015), von Zahnpulpen (TYMOFIYEVA et al. 2013), sowie von Nerven und Nervenkanälen (KRESS et al. 2004, KRASNY et al. 2012) genutzt. Auf Grundlage dieser Untersuchungen wurden in der Pferdemedizin mit Hilfe eines 1,5- und 0,5-Tesla(T)-MRT Aufnahmen von Backenzahnpulpen angefertigt und sowohl physiologische als auch pathologische Veränderungen beschrieben (GERLACH et al. 2013, ILLENBERGER et al. 2013). Kernspintomographen mit einer Feldstärke von 3 T wurden in der Pferdemedizin bereits eingesetzt (CAVALLERI et al. 2013; HONTOIR et al. 2013), jedoch bisher nicht zur Untersuchung dentogener und periodontaler Strukturen. Derart hochauflösende MRT-Geräte besitzen das große Potential, eine diagnostische Aufarbeitung der Zahnweichteilstrukturen in optimaler Bildqualität, ohne den Einsatz von ionisierender Strahlung durchzuführen (KATTI et al. 2011, TYMOFIYEVA et al. 2013). Die Vorteile von Hochfeld-MRTs mit einer Feldstärke von >1 T gegenüber Niedrigfeld-MRTs mit einer Feldstärke von 0,1 bis 0,7 T liegen in einem höheren Signal-Rausch-Verhältnis und einer höheren Magnetfeldhomogenität (WERPY 2007). Die aus hochauflösenden MRT-Bildern gewonnenen Erkenntnisse könnten in der Pferdemedizin zur Beurteilung der Vitalität von Pulpen und zur verbesserten nachfolgenden endodontischen Behandlung bei vorliegender Pulpeninfektion führen (LUNDSTRÖM et al. 2016). Ebenso könnte eine MRT-Untersuchung – wie in der Humanmedizin – die Entscheidungsfindung, ob das periodontale Ligament eines apikal entzündeten Zahnes noch vital ist, maßgeblich beeinflussen. Bei Vorliegen eines gesunden Ligamentes, wäre eine extrakorporale, endodontische Behandlung in Kombination mit einer Reimplantation des betroffenen Backenzahnes möglicherweise eine zahnerhaltende Alternative zur herkömmlichen Extraktion (STOLL u. PEARCE 2017).

Zur Beurteilung pathologischer Befunde sind detaillierte Kenntnisse der physiologischen Darstellung von dentalen und periodontalen Strukturen in der jeweiligen Bildgebungsmodalität von entscheidender Bedeutung. Da die Anatomie, besonders im Bereich der Zähne und der Nasennebenhöhlen, in den verschiedenen Altersklassen variiert, ist das Wissen über altersabhängige physiologische Veränderungen Grundlage für weitere Untersuchungen. Veränderungen betreffen sowohl die Größe der Pulpen, die Lage der Oberkieferbackenzähne im Sinus maxillaris als auch den altersabhängigen Abstand des Canalis infraorbitalis zu den benachbarten Prämolaren und Molaren des Oberkiefers. Während in der Humanmedizin ein altersunabhängiger, direkter Vergleich der Darstellung von Backenzähnen im CT, Cone-Beam-CT und 3-T-MRT durchgeführt wurde (GAUDINO et al. 2011), liegen für equine Patienten keine vergleichbaren Daten vor. Ein Score-basierter Vergleich wurde bislang in Bezug auf die Nasennebenhöhlen (KAMINSKY et al. 2016), nicht aber für dentogene und periodontale Gewebe angefertigt.

Viele Studien zur bildgebenden Diagnostik von Pferdeköpfen werden an Kadavern durchgeführt: Die verwendeten Präparate stammten entweder von Pferden, die unmittelbar postmortal untersucht oder eingefroren und anschließend aufgetaut wurden. Ergebnisse aus 3-T-MRT-Untersuchungen an Pferdebeinen haben gezeigt, dass der Präparatzustand zu einer Veränderung des MR-Signals führen kann: Bei aufgetauten Versuchsobjekten oder Präparaten, die post mortem genutzt wurden, war das MRT-Signal schwächer und die Bilder wenig kontrastreich im Vergleich zu lebenden Patienten (BOLEN et al. 2011). Für die magnetresonanztomographische Darstellung von Pferdezähnen ist bisher unklar, ob der Präparatzustand zu einer Verschlechterung der Bildqualität im MRT führt.

Das Ziel der vorliegenden Studie ist die vergleichende Untersuchung der CT- und MRT-Darstellung von dentogenen, periodontalen und angrenzenden Geweben in verschiedenen Präparatzuständen und an lebenden Patienten. Hierzu werden mittels eines Score-Systems ausgewählte anatomische Strukturen im CT- und 3-T-MRT-Bild durch drei Untersucher bewertet und miteinander verglichen. Dabei erfolgt auch die Untersuchung der altersabhängigen Veränderungen von Oberkieferbackenzähnen und benachbartem Gewebe. Erstmalig soll evaluiert werden, ob der Präparatzustand des Pferdekopfes (in vivo, post mortem, eingefroren-wieder aufgetaut) einen Einfluss auf die MRT-Bildqualität im Bereich des Zahnes hat.

Die Ergebnisse der durchgeführten CT- und MRT-Untersuchungen von physiologischen dentogenen und periodontalen Strukturen können für die Bewertung pathologischer Zahnveränderungen als Grundlage dienen, um eine erfolgreiche Diagnostik und – insofern nötig – eine chirurgische Therapie zu gewährleisten.

2 Publikation I

“Comparison of computed tomography and high-field (3.0 T) magnetic resonance imaging of age-related variances in selected equine maxillary cheek teeth and adjacent tissues“

Christin Schoppe, Maren Hellige, Karl Rohn, Bernhard Ohnesorge, Astrid Bienert-Zeit

BMC Veterinary Research, 13.1 (2017): 280

Akzeptiert am 06.09.2017

2.1 Abstract I

Background

Background: Modern imaging techniques such as computed tomography (CT) and magnetic resonance imaging (MRI) have the advantage of producing images without superimposition. Whilst CT is a well-established technique for dental diagnostics, MRI examinations are rarely used for the evaluation of dental diseases in horses. Regarding equine endodontic therapies which are increasingly implemented, MRI could help to portray changes of the periodontal ligament and display gross pulpar anatomy. Knowledge of age-related changes is essential for diagnosis, as cheek teeth and surrounding structures alter with increasing age. The aim of the present study was to highlight the advantages of CT and MRI regarding age-related changes in selected equine cheek teeth and their adjacent structures.

Results

The CT and MRI appearances of the maxillary 08 s and 09 s and adjacent structures were described by evaluation of post-mortem examinations of nine horses of different ages (Group A: <6 years, B: 6–15 years, C: ≥16 years). Most of the tissues selected were imaged accurately with MRI and CT. Magnetic resonance imaging gives an excellent depiction of soft endo- and periodontal units, and CT of hard dental and bony tissues. Negative correlation between dental age and pulpar sizes was found: 71.3% of the changes in pulp dimensions can be explained by teeth aging. Pulpar sizes ranged from 14.3 to 1.3 mm and were significantly smaller in older horses ($p < 0.05$). A common pulp chamber was present in 33% of the teeth with a mean dental age of 2.25 years. Ninety-four percent of the 08 and 09 alveoli of all groups were in direct contact with the maxillary sinus. An age-related regression was found ($R2 = 0.88$) for the distance between alveoli and the infraorbital canal.

Conclusions

The present study provides information about the dental and periodontal age-related morphology and its visibility using different imaging techniques. These results aid in evaluating diagnostic images and in deciding which is the superior imaging modality for clinical cases.

Schoppe *et al. BMC Veterinary Research* (2017) 13:280
DOI 10.1186/s12917-017-1200-7

BMC Veterinary Research

RESEARCH ARTICLE

Open Access

Comparison of computed tomography and high-field (3.0 T) magnetic resonance imaging of age-related variances in selected equine maxillary cheek teeth and adjacent tissues

Christin Schoppe[1*], Maren Hellige[1†], Karl Rohn[2], Bernhard Ohnesorge[1†] and Astrid Bienert-Zeit[1†]

Abstract

Background: Modern imaging techniques such as computed tomography (CT) and magnetic resonance imaging (MRI) have the advantage of producing images without superimposition. Whilst CT is a well-established technique for dental diagnostics, MRI examinations are rarely used for the evaluation of dental diseases in horses. Regarding equine endodontic therapies which are increasingly implemented, MRI could help to portray changes of the periodontal ligament and display gross pulpar anatomy. Knowledge of age-related changes is essential for diagnosis, as cheek teeth and surrounding structures alter with increasing age. The aim of the present study was to highlight the advantages of CT and MRI regarding age-related changes in selected equine cheek teeth and their adjacent structures.

Results: The CT and MRI appearances of the maxillary 08 s and 09 s and adjacent structures were described by evaluation of post-mortem examinations of nine horses of different ages (Group A: <6 years, B: 6–15 years, C: ≥16 years). Most of the tissues selected were imaged accurately with MRI and CT. Magnetic resonance imaging gives an excellent depiction of soft endo- and periodontal units, and CT of hard dental and bony tissues. Negative correlation between dental age and pulpar sizes was found: 71.3% of the changes in pulp dimensions can be explained by teeth aging. Pulpar sizes ranged from 14.3 to 1.3 mm and were significantly smaller in older horses ($p < 0.05$). A common pulp chamber was present in 33% of the teeth with a mean dental age of 2.25 years. Ninety-four percent of the 08 and 09 alveoli of all groups were in direct contact with the maxillary sinus. An age-related regression was found ($R^2 = 0.88$) for the distance between alveoli and the infraorbital canal.

Conclusions: The present study provides information about the dental and periodontal age-related morphology and its visibility using different imaging techniques. These results aid in evaluating diagnostic images and in deciding which is the superior imaging modality for clinical cases.

Keywords: Horse, Ct, 3 tesla, High-field MRI, Equine maxillary cheek teeth, Age-related variances, Endodontic disease, Periodontal

* Correspondence: Christin.Schoppe@tiho-hannover.de
†Equal contributors
[1]Clinic for Horses, University of Veterinary Medicine Hannover, Foundation, Buenteweg 9, 30559 Hannover, Germany
Full list of author information is available at the end of the article

Background

Radiography has been the primary imaging modality historically for evaluating the heads, paranasal sinuses, and dental and periodontal structures [1]. Modern imaging techniques, such as computed tomography (CT) and magnetic resonance imaging (MRI), offer the benefit of producing three-dimensional (3D) images without superimposition and the possibility of multiplanar reconstructions (MPRs). Both CT and MRI are viable imaging modalities for evaluating the head in human [2–4] and veterinary medicine [5–10]. The well-established CT provides a good spatial resolution and excellent delineation of bony tissue [11], but is limited due to its inability to visualise pulpar and periodontal tissue with detailed resolution [12]. Whilst MRI, with its detailed depiction of soft tissues, has already become a valuable tool for diagnosing oro-dental diseases in humans [13–16], it is rarely used in equine dentistry [17]. Applications reported in the human medical sector include the evaluation of the temporomandibular joint [18, 19] and the nerve channels [20–22], as well as orthodontic utilisation for the assessment of impacted teeth [23], dental pulps or caries diagnosis [13] and apical periodontitis [24]. In accordance with these findings, a 1.5 Tesla MRI was used to portray physiological and pathological cheek teeth [17, 25] and their surrounding structures in equine dentistry [26]. Good MR image acquisition holds enormous potential for radiation-free diagnostic orthodontic workups [27], especially in times when endodontic procedures in conjunction with pulp infection become more common in equine clinical practice [28, 29].

Knowledge of age-related changes and their presentation in 3D imaging modalities is essential for diagnosis and surgical planning since the anatomy of equine hypsodont cheek teeth and neighbouring structures alters with increasing age. While CT, cone-beam CT and 3 Tesla-MRI dental imaging have been directly compared in humans [14], little is known about the dental head-to-head comparison of CT and high-field MRI in different age stages of equine patients.

The purpose of this study was to analyse the visualisation of anatomical landmarks (endo-, periodontal and adjacent structures) and their changes with increasing age in CT and MR images, aiming to highlight the best imaging technique for each structure at different ages. The maxillary 08 s and 09 s were chosen for the examinations because they belong to the cheek teeth with the most frequent pathologies [30].

Methods

Specimens

Nine horses of different breeds (eight warmbloods, one standardbred) were examined to acquire CT and high-field MRI scans of selected equine cheek teeth, their periodontal tissues and adjacent structures. The horses were clinic-owned (TiHo Hannover, Clinic for horses, Germany). The study population was divided into three groups according to age classes: Group A "young" (2–5 years), B "middle-aged" (6–15 years) and C "old to senile" (≥ 16 years), each with n = 3 horses. The age of the selected horses ranged from 2.4 to 21.7 years (median 8.5 years). All horses were subjected to euthanasia on human grounds for non-dental reasons and for health purposes not related to this study. When evaluating the CT and MR images, the Triadan system was used for numbering the cheek teeth [31]. A total of 36 maxillary cheek teeth were examined. The sample pool included eighteen Triadan 08 s and eighteen Triadan 09 s. Owing to eruption time, dental age was determined by subtracting the earliest possible eruption age of the tooth from the age of the horse [32]. An eruption time of 3.5 years for the 08 s and six months for the 09 s was assumed [33]. The population included selected teeth between 0.9 and 20.95 years (median age 7.95 years).

Diagnostic imaging techniques and image evaluation

The imaging processes were carried out at the University of Veterinary Medicine Hannover, Foundation. Images were taken within six hours after euthanasia. The horses were first placed on a stationary CT table in right lateral recumbency and afterwards in dorsal recumbency on a non-stationary MRI table. Dorsal and transversal scan series of the head were acquired (Table 1). The

Table 1 Imaging techniques and settings

Imaging technique	Sequence	Orientation	Matrix	TR (ms)	TE (ms)	ST (mm)
MRI	T1w	3D	1024	8.5	3.9	0.9
	T2w	transverse	1024	4500	80	3.1
	T2w	dorsal	1024	3000	80	4
	PDw	dorsal	1024	6400	30	4
	STIR	transverse	960	8872	30	3
CT	Helical Scan	transverse	1024			1.5

TE echo time, *TR* time to repetition, *ST* slice thickness

orientation for transverse planes was perpendicular and dorsal planes were orientated parallel to the hard palate.

The CT image acquisition was performed with a Brilliance™ CT – Big Bore Oncology Scanner (Philips Medical System, Best, The Netherlands). Images of the head were acquired using a modified dental scanning protocol with 140 kV and 300 mAs. A Philips Achieva™ 3.0TX-Series® was used for MRI acquisition. Surface coils (Philips SENSETM FlexM® and Philips SENSETM FlexL®) were positioned around the region of interest (ROI), between the rostral margin of the facial crest and the orbital cavity. The MRI sequences obtained were: T1 weighted (T1w), T2 weighted (T2w), proton density weighted (PDw) and PD fat-suppressed short-tau-inversion-recovery (STIR) images.

Visualisation and measurements

The CT and MR images were examined and visualisation of the dental and adjacent structures depicted was performed. All examinations were observed by M.H., an experienced radiologist, A.B.-Z., Diplomate EVDC (Equine), and C.S., a trained veterinarian.

Anatomical landmarks (endo-, periodontal and adjacent structures) and their visualisation in CT and MRI were described (Tables 2 and 3).

Special attention was paid to age-related changes. The positional relations between the dental roots and the floor of the paranasal sinuses were described. Pulpar sizes and the positional relation between the dental alveoli and the infraorbital canal (IOC) were portrayed. Each pulp was measured in the mid-tooth section of the cheek tooth selected (Fig. 1a) after each tooth's half-length was determined in the transverse MRI scans. The distance between the dental alveoli and the IOC was portrayed in transverse sections of the CT images. Therefore, the line between the dental alveolus above the pulp position five and the inner border of the IOC was measured (Fig. 1b). Pulpar sizes and the distance between the IOC and the dental alveoli were measured by one of the examiners (C.S.) using easyVET image editor (easy Vet, IFS Informationssysteme GmbH, Hannover, Germany). All measurements were repeated three times under the same conditions and the median value for each pulp or each distance (between the dental alveoli and IOC) was obtained. Moreover, the IOC's shape was assessed.

Statistical analysis

Data were collected on spreadsheets (Excel® 2010, Microsoft® Corporation Redmond, Washington, USA). SAS® software (SAS Institute, Cary North Carolina, USA) was used for statistical analysis and GraphPad Software, Inc.® (La Jolla, California, USA) was used for graphical and statistical representations. Data were tested for normal distribution with the Kolmogorov-Smirnov test. Correlation between dental ages and pulp sizes or alveolar-infraorbital distance was tested by Spearman's rho. A Wilcoxon matched pairs signed rank test was chosen to calculate whether there were significant differences in the pulp sizes or alveolar-infraorbital distances between both sides of the maxilla. The Kruskal-Wallis test and Dunn's multiple comparison test were performed to validate significant differences of pulp dimensions between different age groups. A p value <0.05 was considered statistically significant.

Results

Nine horses were examined immediately after euthanasia. Images of the entire head were acquired by CT

Table 2 Anatomical landmarks of dental structures depicted in MRI and CT referring to imaging advantages

Anatomical region		CT	MRI
Endodontic system	Dental sac	Low attenuated tissue surrounding the reserve crown of growing cheek teeth (transverse)	High signal intensity marked off by the alveolar bone with low signal intensity (T2w and STIR, both transverse)*
	Common pulp cavity	Low attenuated pulp tissue in young maxillary teeth (transverse)	Hyperintense soft tissue, present in young cheek teeth (T2w transverse, PDw dorsal)*
	Pulps	Generally five pulp horns with very low density visible, moderate to well distinguishable from the hyperdense hard dental tissue; Pulp: −350 to 500 HU (transverse)	Generally five hyperintense pulp horns visible Very good distinction from the hard dental tissue (T2w, STIR transverse; PDw dorsal)*, especially in young cheek teeth
Anatomical crown	Tooth root	Isodense root (transverse)*	Isointense: with good alignment of the MRI scan, well demarcated structure (T2w transverse)
Clinical crown	Hard dental tissue, distinction to the oral cavity	Dentine, cement and enamel: distinction possible through the different grade of attenuation; Hyperdense enamel and dentine; slightly less opaque cement (transverse, dorsal)*	Dentine, cement and enamel: signal free, not distinguishable (T2w and STIR transverse) Distinction to the oral cavity not displayable, only visible where tongue with moderate signal intensity is in contact with the dental crown or saliva surrounds the tooth (T2w and STIR, transverse)

*superior imaging modality by comparison of the respective structure in CT and MR images (in brackets: the best imaging modalities and section planes)

Table 3 Anatomical landmarks of periodontal and adjacent structures depicted in MRI and CT referring to imaging advantages

Anatomical region		CT	MRI
Periodontal apparatus	Periodontal space: Periodontal ligament (PDL)	Slightly blurry, isodense gap between hard dental tissue and alveolar bone, moderate distinction from hard dental and bony tissue (transverse, dorsal)	Hyperintense gap, good distinction of hard dental tissue and alveolar bone* (T2w and PDw dorsal)
	Alveolar bone	Hyperdense bone* (transverse, dorsal)	Hypointense cortical bone, marked off by the PDL (T2w and STIR transverse)
Maxillary sinus	Mucosa and cortical bone	Respiratory epithelium: not visible; sinus: air-filled and hypodense; hyperdense thin-walled cortical bone* (transverse)	Signal free cortical bone, delineated by a hyperintense line of mucosa* (T2w and STIR transverse)
IOC	Soft tissue inside and bony structures	Soft tissue inside the canal: low attenuated IOC*: high attenuated, very well distinguishable (transverse)	IOC: signal free, becomes evident through the bright mucosa covering the bony surface of the maxillary sinus (T2w and STIR transverse) Soft tissue*: inhomogeneous, high to moderate signal, intense structures inside: nerve with moderate signal intensity, blood vessels with high or low signal intensity (T2w and STIR transverse)

*superior imaging modality by comparison of the respective structure in CT and MR images (in brackets: the best imaging modalities and section planes)

and MRI settings were adapted to accomplish maximum image quality within a reasonable examination time. The field of view (FOV) in the MRI scans was limited from a transverse line caudal to the 09 maxillary teeth to the rostral end of the 08 upper cheek teeth. It ranged from 180 to 250 mm in dorsally orientated sequences and from 180 to 220 mm in transversely orientated MR images. The duration of the CT scans was short (mean ± SD, 2 ± 5 min); the MRI examination took between 80 and 98 min (mean ± SD, 89 ± 9 min). While dorsally orientated MR image acquisition took between 10 (T2w) and 16 min (PDw), transversal scan duration took between 18 (T2w) and 24 min (STIR). The T1w 3D image scan for MPR lasted approximately 14 min.

Imaging modalities

Anatomical landmarks and imaging advantages comparing CT and MR images regarding the maxillary 08 s and 09 s and their adjacent structures were evaluated and summarised in Tables 2 and 3. Acquisition by CT proved to be the best imaging technique for the representation of the bony non-dental structures (alveolar bone, cortical and cancellous maxillary bone and the IOC). These bony structures were well delineated against the low attenuated soft tissues. Dental enamel, dentine and cementum were also detected as structures of specific density, marked with different greyscales by CT imaging, but the differentiation between them was not as clear as the line between the alveolar bone and the periodontal space due to their different opacity. Regarding

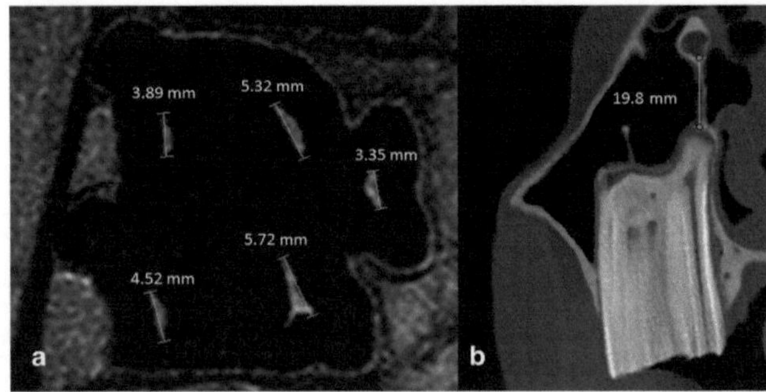

Fig. 1 Measurements of pulpar sizes and distance between alveolus and infraorbital canal (IOC). Pulp dimensions of an 8.5-year-old tooth (108, dorsal plane) in PDw magnetic resonance images (**a**); distance between the alveolar socket and the IOC in a 6-year-old tooth (209, transversal plane) using CT imaging (**b**)

the endodontic system, CT images revealed not only hypodense pulp horns and chambers, but also hypodense funnel-like infundibula. The infundibula could not be detected on the MR images. Marked hypoattenuating stripes were visible adjacent to the infundibula in some of the CT slices, that could be determined as gas. Compared with slightly blurred pulpar CT imaging, the pulp tissue appeared as bright, sharp, hyperintense pulp on the MR images. Other soft tissues, such as the periodontal ligament, respiratory mucosa and the infraorbital nerve, were also better visualised with MRI compared to CT images.

Age-related pulpar sizes

A total of 177 dental pulp measurements were acquired in the upper 08 s and 09 s on dorsally oriented PDw sequences of the MR images. Calculated dental ages for the equine age groups are documented in Table 4. One of the pulp horns was not visible (always pulp 5) in each of three older cheek teeth of the age group B and C (a 10-year-old 209 and a 15.75-year-old 109 and 209). A differentiation between the hypointense hard dental and the normally hyperintense soft pulp tissue could not be performed in any of these MRI sequences because both tissues appeared hypointense in these teeth. Thus, three pulps were excluded from the pulp measurements. The cheek teeth in all other maxillary 08 s and 09 s contained five pulp horns, named pulp 1 to pulp 5 (P1-P5) [34].

Negative correlation between dental age and pulpar dimensions was found ($r = -0.9$). Pulpar size decrease can be explained by dental aging in 71% of the teeth examined: the older the selected cheek teeth of clinically healthy horses became, the smaller the measurements of the pulps (Fig. 2). The sizes of all five pulps varied with age between 14.31 and 1.7 mm for the 08 s and between 14.05 and 1.3 mm for the 09 s. Other pulpar sizes are documented in Table 5. Comparing the pulp horns, pulp 4 was the largest and pulp 5 was the smallest, except for two teeth (Triadan 109) of age group C, where pulp 3 showed the smallest size. Whilst young teeth showed wide pulpar size variations, the range of size differences regarding all pulps became smaller with increasing age; this is visualised by the boxes in Fig. 4. There was no significant difference in the pulpar dimensions between both sides of the maxilla. Significant differences were measurable for the same pulp horn position in all teeth by comparing of the pulp sizes between age group A and C. Pulp 4 also showed a significantly smaller pulpar size between group A and B (Fig. 3). The percentage reduction of mean pulpar size comparing both age classes was between 40 (for all P5) and 56% (for all P4). Pulpar size decrease was smaller considering the same comparisons between the pulps of group B and C, with values ranging from 42 (P3, P2) to 53% (Fig. 4).

Common pulp chambers

The common pulp chambers (CPCs) were evaluated in dorsally orientated PDW sequences of MR images. A CPC was encountered in all cheek teeth of age group A (n = 12). The median dental age for these cheek teeth was 2.025 years and the maximum age was 3.75 years. Cheek teeth of the other age groups showed no CPC.

Positional relations between the maxillary sinus and the dental alveoli

Only two dental alveoli (upper 08 s of age group A) out of all 36 cheek teeth were not located below the maxillary sinus. Positional relations of the structures could be seen in both imaging modalities. Whilst CT images highlighted the bony alveoli, the MRI scans delineated the differentiation of sinus and alveolus through the hyperintense periodontal ligament (PDL) and the sinus mucosa. Eight of the 34 teeth roots located within the boundaries of the sinus cavity were only located below the maxillary sinus with their caudal aspect (08 s), including pulp horn 2 and 4. Fifty percent of these teeth (n = 4) partly contacting the sinus floor contained a CPC. The percentage of teeth that were positioned below the maxillary sinus with all five pulps (n = 26) consists of 08 s (n = 8) and 09 s (n = 18).

Age-related changes of the infraorbital canal and its distance to the dental alveoli

Age-related changes of the IOC were analysed in transverse CT scans showing the positional relation between tooth and bony canal, on the one hand, and the canals' shape, on the other (Fig. 1b). Where the bony IOC was not in contact with the alveolar bone, it was oval-shaped and held in position by a free bone bridge. If the cheek tooth alveolus and the IOC were in close contact, the bony canal was irregularly shaped. Age-related positional changes of the IOC could be measured (Fig. 5). Two teeth roots (both 08 s) of the youngest horse in group A showed direct contact with the IOC, but the distance

Table 4 Calculated dental age of the upper 08 s and 09 s related to the horses' age

Age group	Horses' age in years	Dental age in years: 08 s (n)	Dental age in years: 09 s (n)
A	< 6	< 2.5 (6)	< 5.5 (6)
B	6–15	2.5–12.5 (6)	5.5–15.5 (6)
C	≥ 16	≥ 12.5 (6)	≥ 15.5 (6)

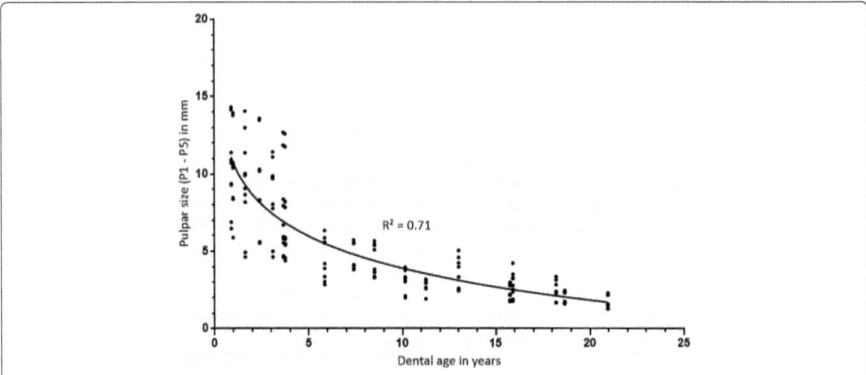

Fig. 2 Semi-logarithmic scale of age-related pulp horn sizes (pulp 1 - pulp 5). Pulps were measured in the mid-tooth section of the maxillary 08 s and 09 s in MR image scans (PDw, dorsal orientation); P1-P5: Pulp horn 1–5

between dental alveolus and IOC increased with dental age. The increase in distance in 88% could be explained by dental aging (R^2 = 0.88). There was no significant difference between both sides of the maxilla.

Discussion

Imaging modalities

As detailed evaluation of the internal and periodontal structures of the cheek teeth is important in deciding on treatment when pulpar or apical infection occurs [35], CT and MR images have been acquired in the present study, aiming to highlight the best imaging modality for each structure. The overview of the two techniques on the same head allows a direct comparison of the potential of both. There are several studies published in equine medicine describing the qualities of CT or MRI for the diagnosis of head pathologies without comparison of both techniques [1–6] and with lower MRI field strength of 0.5 to 1.5 T [17, 25, 36].

The image quality of MRI in the present study was comparable to that of CT, but was better for bony tissues in CT and for soft tissues in MRI. The present results correspond with the findings of Gerlach et al. [36] and Kaminsky et al. [10], who proved the excellent quality of MRI in portraying the dental pulp, PDL, bone-marrow, gingiva, facial soft tissue, sinus mucosa, infraorbital nerve and vessels, owing to a high water content and hydrogen atoms. There was no detectable MRI signal from hard dental tissues, cortical bone, lamina dura and the IOC, whereas these structures could be visualised with good detail resolution by CT acquisition. The best MR image quality in the present study could be achieved with T2w scans for an anatomical overview and with PDw or STIR scans for more detailed images; this is in line with the findings of Kraft and Gavin [37], where T2w, STIR and PDw scans appeared superior to T1w scans in all images evaluated. This is related to the thinner slice thickness of the 3D T1w sequences, which are more often affected by artefacts compared to the transversal and dorsal orientated sequences of the STIR, PD and T2 weighted images with thicker slices.

Compared to low-field tomography, 3.0 T MRI has a higher signal-to-noise ratio, leading to better resolution. Consequently, sequences can be taken with a lower

Table 5 Pulp parameters. MRI measurements were conducted in the mid-tooth section of the upper maxillary 08 s and 09 s. Results show pulp parameters independent of age; n = number of pulp horns; P1-P5: Pulp 1–5

	P1	P2	P3	P4	P5
n	35	36	34	36	36
Mean (mm)	4.89	5.81	5.57	7.32	3.45
s.d. (mm)	2.59	3.26	3.42	4.78	1.65
Median (mm)	3.89	3.75	5.24	5.67	3.17
Range (mm)	7.15	8.76	10.09	13.02	5.30
$x_{min} - x_{max}$ (mm)	2.19–9.34	2.16–0.91	1.3–11.39	1.28–14.31	1.57–6.87

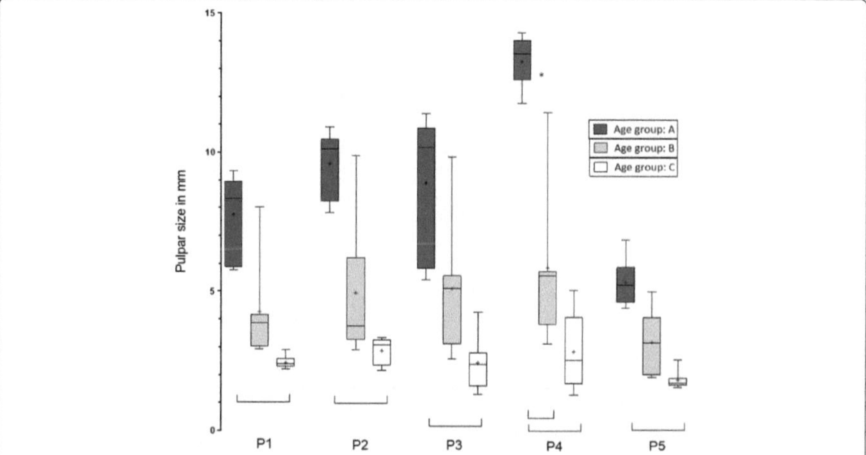

Fig. 3 Pulp measurements of MR images in different age groups. Examination was performed with PDw scans. Horizontal whiskers represent statistically significant differences in size between pulp horns. Boxes represent the interquartile range, vertical whiskers the range and "+" show mean in all boxplots; P1-P5: pulp horn 1–5

recording time and equal image quality compared to low-field MRI. The time for general anaesthesia for patients can be shortened and anaesthesia risk is reduced [38, 39]. The time required for imaging examination in the present study differed greatly between MRI and CT and between the different MRI scans. The CT was around 18 times faster than the entire MRI examination, due to the different imaging techniques and various MRI scans. The long MRI examination times were chosen to acquire images of excellent quality in the current study.

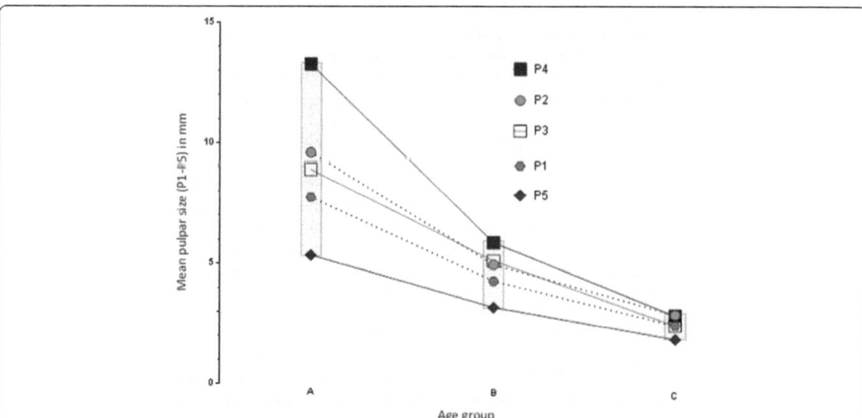

Fig. 4 Different mean reduction of each pulp between different age groups. Mean pulpar sizes presented in different age groups, measured in dorsally orientated MRI scans. Boxes show the range between the mean largest and smallest pulp in each age group. P1 – P5 present pulp horn 1 to pulp horn 5

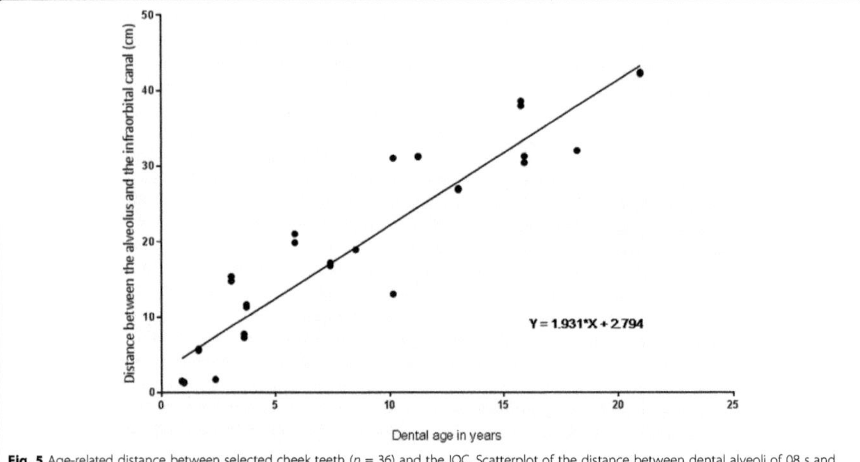

Fig. 5 Age-related distance between selected cheek teeth (n = 36) and the IOC. Scatterplot of the distance between dental alveoli of 08 s and 09 s and the bony IOC measured in transverse CT images

The number of image alignments (transverse, dorsally orientated), resolution, or FOV might be decreased for clinical use to reduce scan time.

Due to the differences in the alignment of teeth [40], the MR images obtained could include artefacts, because image alignment was chosen for the entire skull and not for each tooth. Finally, some pulps and PDLs appeared blurry only due to image angulation (T2w, PDw, STIR) in the current study, that might be mistaken for pathology [17]. The alignment for each tooth would take more time and anaesthetic risk in clinical patients would increase due to longer examination times [41]. T1w images with 3D datasets offer an exception: through MPR, this technique provides the possibility of producing image planes in alignment with the teeth depicted. The time for image acquisition in T1w sequences is prolonged in high-field MRI, as T1 relaxation time is longer [38]. Therefore, a short acquisition time for T1w images, as was applied in the current study, is always accompanied by worse image quality compared to the T2w scans.

In contrast to MRI, CT provides the possibility of different angulations through subsequent MPR and, thus, the potential to evaluate the dental and periodontal changes in alignment of every single tooth. Nevertheless, CT also shows limitations, as it is less useful in identifying early changes in the pulp [42] due to its inability to visualise soft tissues as excellently as T2w or PDw images.

Selected cheek teeth

As described in the literature, dental pathologies such as apical infections, dental fractures and infundibular caries occur predominantly in certain maxillary cheek teeth [43]. The upper 08 s and 09 s show clinical signs most frequently [44–46] and, therefore, they were chosen for examination in the current study.

Selected age groups

The planning of surgical procedures in cases of dental and associated pathologies requires accurate anatomical knowledge and imaging of the cheek teeth and adjacent structures, which are illustrated in this study. Selected equine groups of different ages were chosen in our study as dental changes can be influenced by morphological, functional and mechanical changes, which occur with increasing age [47]: Equine cheek teeth show second dentition and real longitudinal growth up to the horse's age of five years (Group A in the current study); afterwards, the tooth length decreases as teeth are pulled out of the alveoli in the oral direction and the cheek teeth abrasion continues (Group B in our study: 6–15 years) [48, 49]. In older horses (Group C from an age of 15 years onwards), cheek teeth have an exposition to higher dental forces [50]. All these aspects might influence changes of dental structures (e.g. pulpar size) or positional relations between dental and periodontal structures, which were examined in the current study.

Regarding the preselected horse groups (resulting in preselected dental age groups), the present evaluations should be treated with considerable caution: due to the small size of the study population, age distribution and image interpretation might not comply with the entire equine population.

Age-related variances

Common pulp chamber

It can be assumed that CPCs are more common in younger teeth [51], which is in line with the current results, as CPCs were displayed in all teeth evaluated of age group A with a mean dental age of 2.25 years. Dacre et al. [32] located a CPC in teeth with a mean dental age of 2 years (via microscopic examination), whereas Kopke et al. [52] stated the maxillary CPC to be evident in teeth with a mean dental age of 4 years (in high resolution micro-CT scans). While CT [28] and MRI studies [25] observed a CPC in cheek teeth up to 6 years and with micro-CT even up the dental age of 9 years [52] in the present study, the oldest teeth showing a CPC had a dental age of 3.75 years. The difference in results could be caused by the different imaging techniques, imaging settings or measuring techniques. More precise results could probably be achieved by micro-CT scans and histological investigations.

Pulps

Anatomical examinations of the endodontic system by Baker [53] and Dacre et al. [32], as well as MRI [25] and CT [28] imaging studies revealed a general pattern of five pulp horns in the maxillary 08 s and 09 s. These findings comply with the results of the current study, where five main pulp horns (P1-P5) were visible in 33 cheek teeth in the CT and MR images. Three out of all 36 cheek teeth evaluated showed only four pulps.

Contrary to the histological findings of Shaw et al. [54], which displayed an increase in pulpar sizes between 3.5- and 7-year-old teeth, the MRI measurements in the current study showed a continual reduction in size with age. This is consistent with MRI [25] and CT [28] studies demonstrating pulpar reduction with age. The reasons for the pulpar size decrease can be found in age-related physiological pulp modifications. Young equine pulp tissue consists of odontoblasts, connective tissue, nerves, vessels and different cell types. Attachment of secondary dentine is detected and the vital cell number decreases with age [51]. As secondary dentine contains fewer protons, the pulp appears smaller in older equine cheek teeth in the MR images. In the CT images, pulpar tissue does not appears as well delineated as in the MRI scans. Therefore, pulpar existence, dimension and pathologies might not be detected as well with CT imaging as in the MRI scans. Finally, both imaging modalities, CT and MRI, seem to be inferior to histological examinations, which could be the reason for different pulpar dimensions found by Shaw et al. [54] and in imaging studies such as the current one.

In the present study, three pulps could not be measured in the MR images: one pulp was missing completely in each of the 09 s affected (Age group B: $n = 1$; Age group C: $n = 2$). All three teeth with one missing pulp in the MRI scans showed a higher attenuated or a gas spot-filled pulp in the CT scans. Missing pulps have been described by Gasse et al. [55], who carried out research on pulpar changes of the 07 s and 09 s in horses of different ages. Due to the expansion and proliferation of secondary dentine, the number of vital pulps is reduced in horses aged between 15 and 23 years. Fewer hydrogen atoms might result in less pulp detection in MRI scans and secondary dentine might be the explanation for higher attenuated pulps in the CT images of the current study.

Reasons for the absence of pulp in the MRI scans can also be found in pathogenic mechanisms that become visible comparing both imaging techniques. Whilst the MRI showed no causes for the absence of the pulp, CT revealed indications for infundibular and pulpar gas spots that can be interpreted as infundibular caries in the 09 s affecting the pulp horn. While Veraa et al. [56] and Bühler et al. [43] argued that infundibular changes often appear as a singular dental change in CT without significant relationship with pulpitis, Dacre et al. [57] describe infundibular changes which can cause and result in pulpitis, collapsing into the adjacent pulp. Inflammatory cells proliferate as a consequence of the pulpitis, which leads to varied pulp capillary blood flow, arteriovenous anastomoses and ischemic necrosis of the pulp [58]. Finally, this initiates the reduction of pulp size through the production of tertiary dentine [59], decreasing the pulpar visibility in MRI.

Other processes, such as dental trauma [43] or haematogenous pulp infection [47], can result in secondary pulpitis. In addition to pulp stones, histologically referred to as intra-pulpar calcified structures without tubular configuration [32], all these mechanisms can cause decreased vascularity of the pulp itself [60], pulpar infection and destruction, and its inability to be visualised in MR images.

It is described in other studies that the pulps sometimes underwent an occlusal insult, resulting in occlusal necrosis and the production of repairing tertiary dentine, but more apically the horns were vital [57]. This is the reason that all dorsally orientated MRI section planes should be assessed and reviewed for vital pulp tissue.

As a pathologically decreased or missing pulp in MRI does not always allow any conclusions regarding the aetiopathogenesis [17], evaluation of the adjacent structures complemented by CT imaging of bony and hard dental structures and occlusal surfaces is important.

Although all the horses examined appeared to be clinically healthy in terms of their teeth, the missing pulps (MRI) or infundibular gas spots (CT) might be an indicator for the start of dental pathology.

Distance between dental alveoli and the maxillary sinus and the infraorbital canal

Precise anatomical knowledge is essential for the evaluation of dental ascending infections and planning of surgical procedures. If a tooth with apical infection is located within the boundaries of the sinus cavity and induces a sinusitis, treatment of the sinus affected may be indicated [61].

Various declarations referring to the contact between teeth and paranasal sinuses due to inter-individual skull anatomy and age-related variances exist in the literature. While Hillmann [62] revealed that only the last three cheek teeth are in contact with the sinus floor, Dyce et al. [63] mentioned that the last premolar tooth's alveolus is also in contact in young horses. In the current study, 94% of Triadan 08 and 09 showed contact with the maxillary sinus floor. The results obtained correspond to the increased risk of inducing secondary sinusitis in apically infected 08 s and 09 s that is described in literature [64].

Both the one-year-old teeth of Triadan 08 which did not show any contact with the sinus floor had direct contact to the IOC. If the IOC is in close contact with an infected tooth, local bony necrosis could occur due to facilitated expansion of proteolytic enzymes [65]. The distance between the alveoli of 08 s and 09 s and the IOC increased by an average of 1.9 mm per year with age. While a five-year-old tooth shows about 12 mm to the IOC, the distance is measured at 41 mm for a 20-year-old 08 or 09. Infraorbital nerve trauma, associated with neuritis and headshaking, is described as a complication of surgical tooth extraction and sinustomy [66]. Knowledge of the IOC's position, as outlined above, can be essential for the prevention of these complications.

Conclusions

The present study provides information about the dental and periodontal age-related morphology and its visibility via different imaging techniques in clinically healthy horses. Both 3 T MRI and CT provide a valuable tool for the visualisation and detection of dental and adjacent tissues. Both modalities complement each other, because MRI highlights soft dental and adjacent soft tissues and CT depicts hard dental and bony surrounding structures. The results aid in evaluating CT and MR images and in choosing the superior imaging modality. Although MRI is not applied as a routine diagnostic measure in dental pathologies, it provides some advantages that could be used for the detection of pulpar changes or before endodontic procedures. Future investigation is warranted to prove the opportunities of detecting and distinguishing pathological processes comparing MR and CT imaging. Knowledge about the age-related positional relations of the maxillary 08 s and 09 s and their adjacent structures expanded with imaging modalities (MRI, CT) might help to evaluate further clinical cases. False positive or negative results of dental pathologies can be avoided through optimal 3D imaging. Additionally, intra- or postoperative complications, such as IOC damage, can be prevented, as surgical planning is optimised.

Abbreviations

3D: three dimensional; CPC: Common pulp chamber; CT: Computed tomography; e.g.: exempli gratia = for example; FOV: Field of view; IOC: Infraorbital canal; ISS: Inter-slice spacing; min: minutes; MPR: Multiplanar reconstruction; MR: Magnetic resonance; MRI: Magnetic resonance imaging; PDL: Periodontal ligament; PDw: Proton density weighted; pulp 1–5: pulp horn 1–5: P1-P5; ROI: Region of interest; ST: Slice thickness; STIR: Short-time-inversion-recovery; T1w: T1 weighted; T2w: T2 weighted; TE: Echo time; TR: Repetition time; WL: Window level; WW: Window width

Availability of data

The datasets used and analysed during the current study are available from the corresponding author on reasonable request.

Consent to publish

Not applicable.

Authors' contributions

CS designed the study, optimised CT settings, acquired CT and MR data, analysed data, performed measurements and wrote the manuscript. MH supported and supervised the CT and MR data acquisition. AB and MH contributed to the study design, data and image analysis and interpretation of both. AB advised CS in dental nomenclature and preparing the manuscript. BO contributed to the study design. KR proofed the statistical analysis which CS performed. All authors read and approved the final manuscript.

Ethics approval

Studies were approved by the Lower Saxony State Office for Consumer Protection and Food Safety (File reference 33.12-42,502.04.14/1644). The horses were clinic-owned (TiHo Hannover, Clinic for horses, Germany).

Competing interests

None of the authors have any personal or financial relationships which could inappropriately influence or bias the content of this paper.

Publisher's Note

Springer Nature remains neutral with regard to jurisdictional claims in published maps and institutional affiliations.

Author details

[1]Clinic for Horses, University of Veterinary Medicine Hannover, Foundation, Buenteweg 9, 30559 Hannover, Germany. [2]Institute for Veterinary Biometry and Epidemiology, University of Veterinary Medicine Hannover, Foundation, Buenteweg 2, 30559 Hannover, Germany.

Received: 7 September 2016 Accepted: 24 August 2017
Published online: 06 September 2017

References

1. Tremaine WH, Dixon PM. A long-term study of 277 cases of equine sinonasal disease. Part 1: details of horses, historical, clinical and ancillary diagnostic findings. Equine Vet J. 2001;33:274–82.
2. Hirschfelder U, Hirschfelder H. Imaging of the form and structure of the mandible by computed tomography. Fortschr Kieferorthop. 1985;46:138–48.
3. Fuhrmann R, Wehrbein H, Diedrich P: Dreidimensionale computertomographische Darstellung des bezahnten Alveolarkamms Fortschr Kieferorthop 1993;54:91–100.
4. Ordinola-Zapata R, Bramante CM, Duarte MH, Fernandes LMR, Camargo EJ, de Moraes IG, Bernardineli N, Vivan RR, Capelozza ALA, Garcia RB. The influence of cone-beam computed tomography and periapical radiographic evaluation on the assessment of periapical bone destruction in dog's teeth. Oral Surg Oral Med Oral Pathol Oral Radiol Endod. 2011;112:272–9.
5. Moore MP, Gavin PR, Kraft SL, DeHaan CE, Leathers CW, Dorn RV. MR, CT and clinical features from four dogs with nasal tumors involving the rostral cerebrum. Vet Radiol. 1991;32:19–25.
6. Arencibia A, Vazquez JM, Jaber R, Gil F, Ramirez JA, Rivero M. Magnetic resonance imaging and cross sectional anatomy of the normal equine sinuses and nasal passages. Vet Radiol Ultrasound. 2000;41:313–9.
7. Tucker RL, Sampson SN. Magnetic resonance imaging protocols for the horse. Clin Tech Equine Pract. 2007;6:2–15.
8. Bienert A. Digitalradiographische, computertomographische und mikrobiologische Untersuchungen bei Backenzahnerkrankungen des Pferdes. Dissertation. Hannover: Tierärztliche Hochschule Hannover; 2002.
9. Gerlach K, Gerhards H. Magnetresonanztomographische Merkmale von Zubildungen im Bereich der Nase, Nasennebenhöhlen und der angrenzenden Knochen: retrospektive Analyse von 33 Pferden. Pferdeheilkunde. 2008;24:565–76.
10. Kaminsky J, Bienert-Zeit A, Hellige M, Ohnesorge B. 3 tesla magnetic resonance imaging of the equine nasal cavities, paranasal sinuses and adjacent anatomical structures in 13 healthy horses. Pferdeheilkunde. 2014; 29:183–8.
11. Barbee DD, Allen JR, Gavin PR. Computed tomography in horses. Vet Radiol. 1987;28:144–51.
12. Newton CW, Hoen MM, Goodis HE, Johnson BR, McClanahan SB. Identify and determine the metrics, hierarchy, and predictive value of all the parameters and/or methods used during endodontic diagnosis. J Endod. 2009;35:1635–44.
13. Tymofiyeva O, Boldt J, Rottner K, Schmid F, Richter EJ, Jakob PM. High-resolution 3D magnetic resonance imaging and quantification of carious lesions and dental pulp in vivo. MAGMA. 2009;22:365–74.
14. Gaudino C, Cosgarea R, Heiland S, Csernus R, Beomonte Zobel B, Pham M, Kim TS, Bendszus M, Rohde S. MR-imaging of teeth and periodontal apparatus: an experimental study comparing high-resolution MRI with MDCT and CBCT. Eur Radiol. 2011;21:2575–83.
15. Idiyatullin D, Corum C, Moeller S, Prasad HS, Garwood M, Nixdorf DR. Dental MRI: making the invisible visible. J Endod. 2011;37:745–52.
16. Assaf AT, Zrnc TA, Remus CC, Schönfeld M, Habermann CR, Riecke B, Friedrich RE, Fiehler J, Heiland M, Sedlacik J. Evaluation of four different optimized magnetic-resonance-imaging sequences for visualization of dental and maxillo-mandibular structures at 3 T. J Craniomaxillofac Surg. 2014;42:1356–63.
17. Gerlach K, Ludewig E, Brehm W, Gerhards H, Delling U. Magnetic resonance imaging of pulp in normal and diseased equine cheek teeth. Vet Radiol Ultrasound. 2013;54:48–53.
18. Şener S, Akgünlü F. MRI characteristics of anterior disc displacement with and without reduction. Dentomaxillofac Radiol. 2014;33:245–52.
19. Orhan K, Nishiyama H, Tadashi S, Shumei M, Furukawa S. MR of 2270 TMJs: prevalence of radiographic presence of otomastoiditis in temporomandibular joint disorders. Eur J Radiol. 2005;55:102–7.
20. Nasel C, Gahleitner A, Breitenseher M, Czerny C, Glaser C, Solar P, Imho H. Localization of the mandibular neurovascular bundle using dental magnetic resonance imaging. Dentomaxillofac Radiol. 1998;27:305–7.
21. Kress B, Gottschalk A, Anders L, Stippich C, Palm F, Bähren W, Sartor K. High-resolution dental magnetic resonance imaging of inferior alveolar nerve responses to the extraction of third molars. Eur Radiol. 2004;14:1416–20.
22. Krasny A, Krasny N, Prescher A. Anatomic variations of neural canal structures of the mandible observed by 3-tesla magnetic resonance imaging. J Comput Assist Tomogr. 2012;36:150–3.
23. Tymofiyeva O, Proff PC, Rottner K, Düring M, Jakob PM, Richter EJ. Diagnosis of dental abnormalities in children using 3-dimensional magnetic resonance imaging. J Oral Maxillofac Surg 2013;71:1159–1169.
24. Geibel M, Schreiber E, Bracher A, Hell E, Ulrici J, Sailer L, Ozpeynirci Y, Rasche V. Assessment of apical periodontitis by MRI: a feasibility study. RöFo. 2015; 187:269–75.
25. Illenberger N, Brehm W, Ludewig E, Gerlach K. Darstellung altersabhängiger Veränderungen der Zahnpulpen ausgewählter Oberkieferbackenzähne des Pferdes mittels Magnetresonanztomographie. Pferdeheilkunde. 2013;29:183–8.
26. Gerlach K, Flatz K, Brehm W, Seeger J. Klinische Anatomie des Gesichtsbereiches des Pferdes in der Magnetresonanztomographie. Pferdeheilkunde. 2009;25:45–52.
27. Katti G, Ara SA, Shireen A. Magnetic resonance imaging (MRI)–a review. Internat J Dent Clin. 2011;3:65–70.
28. Windley Z, Weller R, Tremaine WH, Perkins JD. Two- and three-dimensional computed tomographic anatomy of the enamel, infundibulae and pulp of 126 equine cheek teeth. Part 2: findings in teeth with macroscopic occlusal or computed tomographic lesions. Equine Vet J. 2009;41:441–7.
29. Lundström T, Wattle O. Description of a technique for orthograde endodontic treatment of equine cheek teeth with apical infections. Equine Vet Educ. 2016;28:1–12.
30. Baker GJ. Dental decay and endodontic disease. In: Baker GJ, Easley J, editors. Equine dentistry. 1st ed. London: WB Saunders; 1999. p. 79–84.
31. Floyd M. The modified Triadan system: nomenclature for veterinary dentistry. J Vet Dent. 1991;8:18–9.
32. Dacre IT, Kempson S, Dixon PM. Pathological studies of cheek teeth apical infections in the horse. 1. Normal endodontic anatomy and dentinal structure of equine cheek teeth. Vet J. 2008;178:311–20.
33. Nickel R, Schummer A, Seiferle E. Eingeweide. Lehrbuch der Anatomie der Haustiere. 2nd ed. Stuttgart: Georg Thieme Verlag; 2004.
34. Du Toit N, Kempson SA, Dixon PM: Donkey dental anatomy. Part 1: gross and computed axial tomography examinations. Vet J 2008, 176:338–344.
35. Manso-Díaz G, García-López JM, Maranda L, Taeymans O. The role of head computed tomography in equine practice. Equine Vet Educ. 2015;27:136–45.
36. Gerlach K, Brehm W, Gerhards W, Ludewig E. Diagnostik von Erkrankungen der Backenzähne des Pferdes mittels Magnetresonanztomographie. Pferdeheilkunde. 2011;27:711–8.
37. Kraft SL, Gavin P. Physical principles and technical considerations for equine computed tomography and magnetic resonance imaging. Vet Clin North Am Equine Pract. 2001;17:115–30.
38. Merkle EM, Dale BM. Abdominal MRI at 3.0 T: the basics revisited. AJR Am J Roentgenol. 2006;186:1524–32.
39. Weishaupt D, Köchli VD, Marincek B. And Borut Marincek. Wie funktioniert MRI. 6th ed. Heidelberg: Springer Medizin Verlag; 2009.
40. Dixon PM. Du ToitN: dental anatomy. In: Easley J, Dixon PM, Schumacher J, editors. Equine dentistry. 3rd ed. Philadelphia: WB Saunders; 2011. p. 25–48.
41. Johnston GM, Eastment JK, Wood JLN, Taylor PM. The confidential enquiry into perioperative equine fatalities (CEPEF): mortality results of phases 1 and 2. Vet Anaesth Analg. 2002;29:159–70.
42. Casey MB, Pearson GR, Perkins JD, Tremaine WH. Gross, computed tomographic and histological findings in mandibular cheek teeth extracted from horses with clinical signs of pulpitis due to apical infection. Equine Vet J. 2015;47:557–67.
43. Bühler M, Fürst A, Lewis FI, Kummer M, Ohlerth S. Computed tomographic features of apical infection of equine maxillary cheek teeth: a retrospective study of 49 horses. Equine Vet J. 2014;46:468–73.
44. Staszyk C, Lehmann F, Bienert A, Ludwig K, Gasse H. Measurement of masticatory forces in the horse. Pferdeheilkunde. 2006;22:12–6.
45. Dacre I, Kempson S, Dixon PM. Equine idiopathic cheek teeth fractures. Part 1: pathological studies on 35 fractured cheek teeth. Equine Vet J. 2007;39:310–8.
46. Staszyk C. Anatomie. In: Vogt C, editor. Lehrbuch der Zahnheilkunde beim Pferd. 1st ed. Stuttgart: Schattauer; 2011. p. 1–29.
47. Dixon PM, Dacre I. A review of equine dental disorders. Vet J. 2005;169:165–87.
48. Becker E: Zähne. In: Joest E, editors. Handbuch der speziellen pathologischen Anatomie der Haustiere. Berlin-Hamburg: Paul Parey Verlag; 1970. p. 83–278.
49. Budras KD, Röck S. Atlas der Anatomie des Pferdes. 3rd ed. Schlütersche: Hannover; 1997.

50. Cordes V, Gardemin M, Lüpke M, Seifert H, Borchers L, Staszyk C. Finite element analysis in 3-D models of equine cheek teeth. Vet J. 2012;193:391–6.
51. Kirkland KD, Baker GJ, Manfra Marretta S, Eurell JAC, Losonsky JM. Effects of aging on the endodontic system, reserve crown, and the roots of equine mandibular cheek teeth. Am J Vet Res. 1997;57:8–31.
52. Kopke S, Angrisani N, Staszyk C. The dental cavities of equine cheek teeth: three-dimensional reconstructions based on high resolution micro-computed tomography. BMC Vet Res. 2012;8:1–16.
53. Baker GJ. Some aspects of equine dental disease. Equine Vet J. 1970;2:105–10.
54. Shaw DJ, Dacre IT, Dixon PM. Pathological studies of cheek teeth apical infections in the horse: 2. Quantitative measurements in normal equine dentine. Vet J. 2008;178:321–32.
55. Gasse H, Westenberger W, Staszyk C. The endodontic system of equine cheek teeth: a reexamination of pulp horns and root canals in view of age-related physiological differences. Pferdeheilkunde. 2004;20:13–8.
56. Veraa S, Voorhout G, Klein WR. Computed tomography of the upper cheek teeth in horses with infundibular changes and apical infection. Equine Vet J. 2009;41:872–6.
57. Dacre I, Kempson S, Dixon PM. Pathological studies of cheek teeth apical infections in the horse: 5. Aetiopathological findings in 57 apically infected maxillary cheek teeth and histological and ultrastructural findings. Vet J. 2008;178:352–63.
58. Syngcuk K, Trowbridge H. Pulpal reaction to caries and dental procedures. In: Cohen S, Burns RC, editors. Pathways of the pulp. 7th ed. St. Louis: Mosby; 1998. p. 414–33.
59. Dacre IT, Shaw DJ, Dixon PM. Pathological studies of cheek teeth apical infections in the horse: 3. Quantitative measurements of dentine in apically infected cheek teeth. Vet J. 2008;178:333–40.
60. Torneck CD. Dentin-pulp complex. In: Ten Cate AR, editor. Oral histology: development, structure and function. 5th ed. St. Louis: Mosby; 1985. p. 146–82.
61. Brinkschulte M. Morphologische Untersuchung der Apertura nasomaxillaris des Pferdes sowie deren Verzweigung in die Nasennebenhöhlen unter Anwendung dreidimensionaler Rekonstruktion computertomographischer Schnittbildserien. Doctoral dissertation. Hannover: Tierärztliche Hochschule Hannover; 2012.
62. Hillmann DJ. Skull. In: Getty R, editor. Sissons and Grossman's the anatomy of the domestic animals. 5th ed. Philadelphia: WB Saunders; 1975. p. 335–6.
63. Dyce KM, Sack WO, Wensing CJG. Anatomie der Haustiere – Lehrbuch für Studium und Praxis. Stuttgart: Enke; 1997.
64. Henninger W, Frame EM, Willmann M, Simhofer H, Malleczek D, Kneissl SM. CT features of alveolitis and sinusitis in horses. Vet Radiol Ultrasound. 2003; 44:269–76.
65. Waguespack RW, Burba DJ, Moore RM. Surgical site infection and the use of antimicrobials. In: Auer JA, Stick JA, editors. Equine surgery. 3rd ed. Philadelphia: Saunders; 2006. p. 70–87.
66. Gilsenan WF, Getman LM, Parente EJ, Johnson AL. Headshaking in 5 horses after paranasal sinus surgery. Vet Surg. 2014;43:678–84.

3 Publikation II

"Magnetic resonance imaging and computed tomography of equine cheek teeth and adjacent structures: comparative study of image quality in horses in vivo, post-mortem and frozen-thawed"

Christin Röttiger. Maren Hellige, Bernhard Ohnesorge, Astrid Bienert-Zeit

Acta Veterinaria Scandinavica, 61 (2019)

Akzeptiert am 02.12.2019

3.1 Abstract II

Background: The use of cadavers for radiology research methodologies which involve subjective image quality evaluation of anatomical criteria is well documented. The purpose of the current method comparison study was to evaluate the image quality of dental and adjacent structures in computed tomography (CT) and high-field (3 Tesla) magnetic resonance (MR) images in cadaveric heads, based on an objective four-point rating scale. Whilst CT is a well-established technique, MR examinations are rarely used for dental diagnostics in horses. The use of a grading system in the present study allowed for an objective assessment of CT and MR imaging advantages in portraying equine cheek teeth. As imaging is commonly performed with cadaveric or frozen and thawed heads for dental research investigations, the second objective of the current study was to quantify the impact of specimens' conditions (alive, postmortem, frozen-thawed) on the image quality in CT and MRI.

Results: CT and MR images of nine horses, focused on the maxillary premolar 08s and molar 09s, were acquired postmortem (Group B). Three observers scored dental, periodontal and adjacent tissues. Results showed that MR sequences gave an excellent depiction of endo- and periodontal structures, whereas CT produced high-quality images of the hard tooth and bony tissues. In three of these nine horses, additional CT and MR imaging (MRI) was performed in vivo (Group A) and frozen-thawed (Group C) to specify the best specimens` condition for further research. Assessing the impact of specimens' conditions on image quality, specific soft tissues of the maxillary 08 and 09 including adjacent structures (pulps, mucosa of the maxillary sinuses, the periodontal ligament, the soft tissue inside the infraorbital canal) were graded in group A and C and analysed for significant differences within CT and MR modalities in comparison to group B. Results showed that MRI scores of the equine dental structures in vivo were superior compared to the postmortem and the frozen-thawed condition.

Conclusions: The capacity of CT and MRI to evaluate dental and periodontal tissues and the advantage of MRI regarding the image quality in horses alive has been shown based on an objective rating scale.

Röttiger *et al. Acta Vet Scand* (2019) 61:62
https://doi.org/10.1186/s13028-019-0495-8

Acta Veterinaria Scandinavica

RESEARCH Open Access

Magnetic resonance imaging and computed tomography of equine cheek teeth and adjacent structures: comparative study of image quality in horses in vivo, post-mortem and frozen-thawed

Christin Röttiger*, Maren Hellige, Bernhard Ohnesorge and Astrid Bienert-Zeit

Abstract

Background: The use of cadavers for radiology research methodologies involving subjective image quality evaluation of anatomical criteria is well-documented. The purpose of this method comparison study was to evaluate the image quality of dental and adjacent structures in computed tomography (CT) and high-field (3 T) magnetic resonance (MR) images in cadaveric heads, based on an objective four-point rating scale. Whilst CT is a well-established technique, MR imaging (MRI) is rarely used for equine dental diagnostics. The use of a grading system in this study allowed an objective assessment of CT and MRI advantages in portraying equine cheek teeth. As imaging is commonly performed with cadaveric or frozen and thawed heads for dental research investigations, the second objective was to quantify the impact of the specimens' conditions (in vivo, post-mortem, frozen-thawed) on the image quality in CT and MRI.

Results: The CT and MR images of nine horses, focused on the maxillary premolar 08s and molar 09s, were acquired post-mortem (Group A). Three observers scored the dental and adjacent tissues. Results showed that MR sequences gave an excellent depiction of endo- and periodontal structures, whereas CT produced high-quality images of the hard tooth and bony tissues. Additional CT and MRI was performed in vivo (Group B) and frozen-thawed (Group C) in three of these nine horses to specify the condition of the best specimens for further research. Assessing the impact of the specimens' conditions on image quality, specific soft tissues of the maxillary 08s and 09s including adjacent structures (pulps, mucosa of the maxillary sinuses, periodontal ligament, soft tissue inside the infraorbital canal) were graded in group B and C and analysed for significant differences within CT and MR modalities in comparison to group A. Results showed that MRI scores in vivo were superior to the post-mortem and frozen-thawed condition.

Conclusions: On comparing the imaging performance of CT and MRI, both techniques show a huge potential for application in equine dentistry. Further studies are needed to assess the clinical suitability of MRI. For further research investigations it must be considered, that the best MR image quality is provided in live horses.

Keywords: Cheek teeth, CT, Equine dental imaging, High-field magnetic resonance imaging, Horse, Scoring system, Periodontal ligament

*Correspondence: Christin.Roettiger@tiho-hannover.de
Clinic for Horses, University of Veterinary Medicine Hannover, Foundation, Bünteweg 9, 30559 Hannover, Germany

Background

The imaging of equine cheek teeth pathologies, such as apical periodontitis [1], pulpitis [2], infundibular caries [3] or ascending infections [4], has been expanded substantially. Although the clinical dental examination is always the basic start, supplementary imaging might be necessary to make a diagnosis [1]. Therefore, knowledge of the physiological depiction of dental, periodontal and adjacent structures in different imaging modalities is essential to obtain accurate diagnoses.

Radiography has always been the primary, established and most widely used standard for dental imaging in horses when comparing different imaging modalities [5]. Diagnostic imaging procedures added more recently, such as computed tomography (CT) and magnetic resonance imaging (MRI), are characterized by high tissue contrast and the possibility of multiplanar or three-dimensional reconstructions without superimposition [6, 7]. Whereas CT has already been established for diagnosing equine dental pathologies [3, 8, 9], the possibilities of dental MRI diagnoses are rarely used in equine dentistry. The MRI has the potential to produce images with excellent detail of dental soft tissues [2, 10]. Regarding clinical patients, MRI might help to assess the vitality of pulp tissue. The study of [2] showed that evaluation of dental pulp in equine cheek teeth is possible using MRI, as pulp with a blurred or enlarged MR signal was considered diseased. With the information regarding which pulp horn is vital or necrotic, endodontic treatment might be more accurate and purposeful. If it remains unclear (after clinical, radiographic and CT examination) whether the periodontal ligament (PDL) is involved in the pathological dental progress, MRI might help to evaluate the vitality of the PDL due to the different intensities depicted in MRI [11]. Endodontic treatment [12] or replantation [13, 14] of apically infected cheek teeth might be a promising alternative to conventional tooth extractions in teeth with a vital PDL. Recent studies compared CT and 3.0 T dental MRI quantitatively in horses, aiming to highlight the best imaging technique for each structure [10]. Qualitative head-to-head comparisons of CT and different MRI protocols, based on a scoring system, have already been performed in human medicine [15]. The general differences between CT and MRI in dental imaging are widely reported in equine medicine, but a rating scale for detailed, more objective results has not yet been used.

Many research investigations have been conducted as there is a need for a better understanding of the pathogenesis of dental diseases. Most of these research examinations have been performed with cadaveric heads and some imaging procedures are performed on frozen and thawed heads. A decrease of the magnetic resonance (MR) signal was described for equine limbs when evaluating defined structures immediately post-mortem and frozen-thawed [16]. Regarding equine dental imaging, there is currently a lack of information about whether the image quality suffers in equine heads post-mortem or frozen-thawed.

The purpose of the current study was to evaluate the general image quality and visibility of dental, periodontal and adjacent structures in CT and different high-field MRI sequences based on a four-point grading scale in cadaveric heads. Another aim was to assess the impact of the condition of the specimens (horses alive, heads post-mortem or frozen-thawed) on the CT and MRI quality and detailed representation of the structures mentioned above. The authors hypothesize that the image quality may achieve the same results in CT images in all groups, but MRI scores may achieve better results regarding the image quality for the dental and periodontal tissues in horses alive compared to those post-mortem or frozen-thawed.

Methods

Specimens and study design

Nine Warmblood horses were prospectively chosen to undergo CT and high-field MRI to display selected maxillary cheek teeth, their periodontal tissues and adjacent structures. Figure 1 illustrates how the method comparison study was conducted. All horses examined post-mortem (group A, n = 9) underwent a CT and MRI acquisition within four hours after euthanasia. The population of group A consisted of five mares and four geldings with a median age of 8.2 years (2.3 to 22.1 years). All horses were owned by the clinic (University of Veterinary Medicine Hannover, Clinic for Horses, Germany) and humanely put down for reasons unrelated to the study. One of the authors (ABZ) decided on the inclusion of each subject: none of the horses had a known history or clinical signs of a paranasal sinus or dental disease. Any clinical signs of dental (e.g. abnormal feed intake or quidding) or sinus disease (e.g. nasal discharge) resulted in exclusion.

The current examinations were linked to another research study. All nine horses examined post-mortem were derived from the other scientific survey. A part of the study population (group B) has also been examined in vivo. The CT and MR examinations in vivo were only possible for three of the nine horses in the present trial due to the experimental set-up of the linked study. The median age of this study population (n = 3) was 9.1 years. The horses of group B were euthanized within 14 to 16 days after CT and MRI acquisition under general anaesthesia. The heads of these horses were harvested post-mortem at the atlanto-occipital joint and frozen (− 20 °C) for 2 weeks (group C, n = 3). The CT and MRI

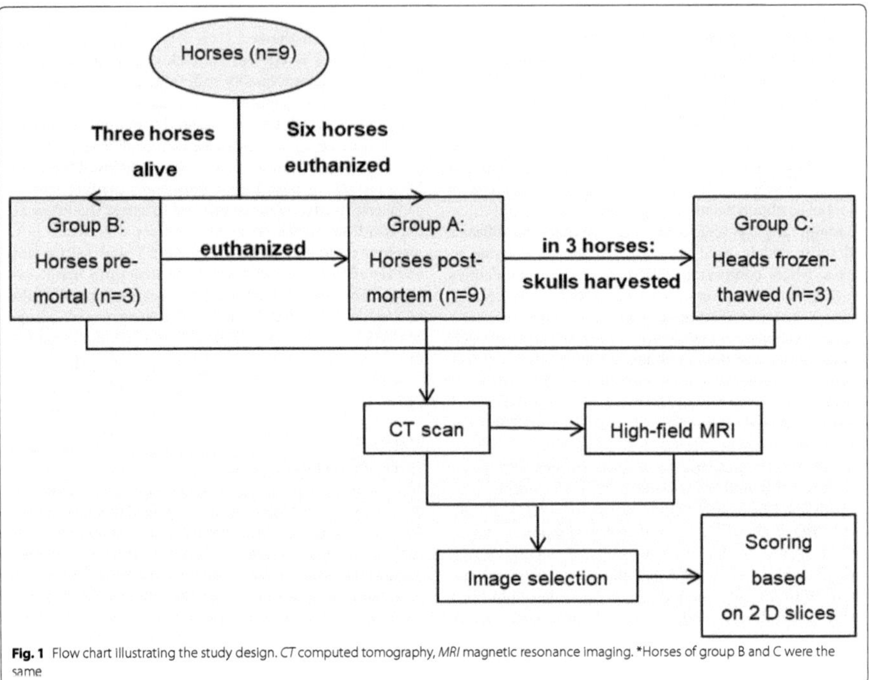

Fig. 1 Flow chart illustrating the study design. *CT* computed tomography, *MRI* magnetic resonance imaging. *Horses of group B and C were the same

datasets were acquired a third time within 48 h after the heads were thawed. The heads were warmed up to a core temperature of 15 °C to prevent imaging artefacts that might occur due to frozen tissues. When assessing the CT and MR images, the modified Triadan system was used for numbering the maxillary cheek teeth [17]. A total of 36 cheek teeth were examined. The images evaluated included 18 maxillary Triadan 08s and 18 maxillary Triadan 09s.

Imaging

The examinations were acquired at the University of Veterinary Medicine Hannover, Foundation. All groups were examined with the same imaging settings for the CT and MRI acquisition. The horses or heads were first positioned on a stationary CT table in right lateral recumbency and, subsequently, in dorsal recumbency on a non-stationary MRI table. Dorsal and transversal slices of the heads were assessed. All CT scans were performed using a 16-line Brilliance™ CT—Big Bore Oncology Scanner (Philips Medical System, Best, The Netherlands). The technical settings were 140 kV, 300 mAs, 1.5 mm collimation, a pitch of 0.9 and a reconstruction matrix of 1024 × 1024. Transverse image series, created with multiplanar reconstruction (MPR), were orientated in alignment with the teeth depicted. The MRI were obtained using a Philips Achieva™ 3.0TX-Series® MRI. Surface coils (Philips SENSETM FlexM® and Philips SENSETM FlexL®) were positioned around the region of interest, between the rostral margin of the facial crest and the orbital region. The MRI scans acquired were: T1 weighted (T1w), T2 weighted (T2w), proton-density weighted (PDw) and PDw fat-suppressed spectral attenuated inversion recovery (PD SPAIR). An additional table file shows this in more detail (see Additional file 1).

Table 1 Evaluated structures depicted in imaging techniques and image alignments

Structures evaluated	Imaging techniques (alignments)
Dental tissues	
Pulp	CT (MPR), PDw (dorsal), PD SPAIR (transverse), T2w (transverse, dorsal)
Common pulp chamber	CT (MPR), PDw (dorsal), T2w (dorsal)
Dental hard tissues (enamel, dentine and cementum), alveolar part	CT (MPR), PDw (dorsal), PD SPAIR (transverse), T2w (transverse, dorsal
Dental hard tissues (enamel, dentine and cementum), clinical crown	CT (MPR), PD SPAIR (transverse), T2w (transverse, dorsal)
Periodontal tissues	
Periodontal ligament	CT (MPR), PDw (dorsal), PD SPAIR (transverse), T2w (transverse, dorsal)
Lamina dura of the maxillary bone	CT (MPR), PDw (dorsal), PD SPAIR (transverse), T2w (transverse, dorsal)
Adjacent tissues	
Cortical bone of the maxillary sinus	CT (MPR), PDw (dorsal), PD SPAIR (transverse), T2w (transverse, dorsal)
Mucosa of the maxillary sinus	CT (MPR), PDw (dorsal), PD SPAIR (transverse), T2w (transverse, dorsal)
Bony infra-orbital canal	CT (MPR), PD SPAIR (transverse), T2w (transverse)
Soft tissue inside the infra-orbital canal	CT (MPR), PD SPAIR (transverse), T2w (transverse)

CT computed tomography, *MPR* multiplanar reconstruction, *PDw* proton-density weighted, *PD SPAIR* proton-density weighted spectral attenuated inversion recovery, *T2w* T2 weighted

Images analyses

After image acquisition, CT and MRI slices from different planes of the cheek teeth and adjacent structures were chosen (Table 1). Three slices through each of the maxillary 08s and 09s were chosen in a dorsal and transversal orientation in the CT, T2w, PDw and PD SPAIR scans.

Predefined anatomical landmarks were used to ensure comparability of the selected slices. The dorsally orientated slice in the mid-tooth section of the maxillary teeth, for example, was selected after each tooth's half-length was determined in the transverse scans. Each structure visible in this slice was scored. Data were exported in DICOM format to easyIMAGE software (easyVet, IFS Informationssysteme GmbH, Langenhagen, Germany). Images were analysed and evaluated on a 19″ flat DICOM certified TFT display (EIZO FlexScan MX190S; EIZO Europe GmbH, Mönchengladbach, Germany).

The images chosen were evaluated independently by three experienced veterinarians (MH, an experienced radiologist and resident in the European College of Veterinary Diagnostic Imaging; ABZ, a board-certified specialist in equine dentistry and CR, a-trained veterinarian). The CT and MR images were graded separately, and information concerning the specimen's condition were hidden. A modified four-point rating scale was used by each observer to analyse the image quality (Table 2), as described in several human and veterinary studies evaluating imaging techniques [15, 18, 19]. Additionally, the visibility and the differentiation (contours and tissue distinction) of specific dental, periodontal and adjacent structures were graded (Tables 1 and 3). The examiners could adjust the window width and level individually.

Statistical analysis

Data were collected on spreadsheets (Excel® 2010, Microsoft® Corporation Redmond, Washington, USA). SAS® software (SAS Institute, Cary, NC, USA) was used for the statistical analyses. GraphPad Software, Inc.® (La Jolla, CA, USA) was chosen for the graphical and statistical representations. Data were tested for normal distribution with Kolmogorov–Smirnov tests and analysed with a

Table 2 Modified scoring system for image quality parameters, according to [18]

Score	Image noise	Image sharpness	Image contrast
0 Poor quality...	... with very high image noise level or image quality highly impaired due to image noise	... due to low image sharpness	... due to low image contrast
1 Moderate quality ...	... with high image noise level or image quality impaired due to image noise	... due to moderate image sharpness	... due to moderate image contrast
2 Satisfactory quality ...	... with moderate image noise level or image quality slightly impaired due to image noise	... due to satisfactory image sharpness	... due to satisfactory image contrast
3 Good quality ...	... without image noise or image quality impaired due to image noise	... due to high image sharpness	... due to high image contrast

Table 3 Modified scoring system for the visibility/distinction of anatomical structures, according to [18]

Score	Criteria for visibility of anatomical structures: Structures ...	Criteria for tissue or contour distinction (differentiation): Contours.../Tissues...
0	... not visible	... not distinguishable
1	... poorly visible, but detectable; identified by its location and signal intensity, not by margins, shape or size	... distinguishable, but often blurred
2	... clearly identified by its location, signal intensity and shape, margins not clearly delineated	... well distinguished and seldom blurred
3	... very well visualized and clearly delineated by location, shape, signal intensity/density, size and margins	... very well distinguished and sharply defined

non-parametric statistical test (Friedman test). Wilcoxon matched pairs signed-rank tests were applied to calculate significant differences between CT and MRI scores. An adjusted α^* was assessed using Bonferroni's procedure to maintain study-related errors. Therefore, each individual hypothesis was tested at a significance level of α/m where $\alpha$is the overall alpha level (0.05) desired and m is the number of hypotheses. The inter-observer agreement was analysed using McNemar-Boker tests and the Cohen's kappa coefficient was calculated.

Results

The CT images, PDw, PD SPAIR and T2w sequences were included in the study. The three-dimensional T1w scans were excluded because the quality was not good enough for further evaluation. The field of view in the MRI scans ranged from 180 to 250 mm in dorsally orientated sequences and from 160 to 220 mm in transversely orientated MRI for all groups. A total of 1080 images were evaluated, and 14,040 parameters were graded by all observers (8424 parameters in group A; 2808 parameters in each of group B and C).

Image quality, structures' visibility and MRI/CT differentiation in horses post-mortem (group A)

Image quality parameters and scores for the visibility of dental (Fig. 2), periodontal (Fig. 3) and adjacent structures (Fig. 4) were analysed. The structures evaluated are depicted in Fig. 5.

The CT scores for image noise (median of 2.66) were significantly superior ($P<0.05$) than the MRI scores for PDw (median of 2.13), PD SPAIR and T2w (median of 2.33) images (P (CT vs. PDw) $=0.0052$; P (CT vs. PD SPAIR) $=0.0014$; P (CT vs. T2w) <0.0001). The CT scores for image sharpness (median of 2.66) showed significantly better results than the MRI scores for PDw, PD SPAIR and T2w images (medians of 2.33) (P (CT vs. PDw) $=0.0019$; P (CT vs. PD SPAIR) $=0.0027$; P (CT vs. T2w) $=0.0008$). Image contrast was graded with a median score of 3 for all imaging techniques acquired

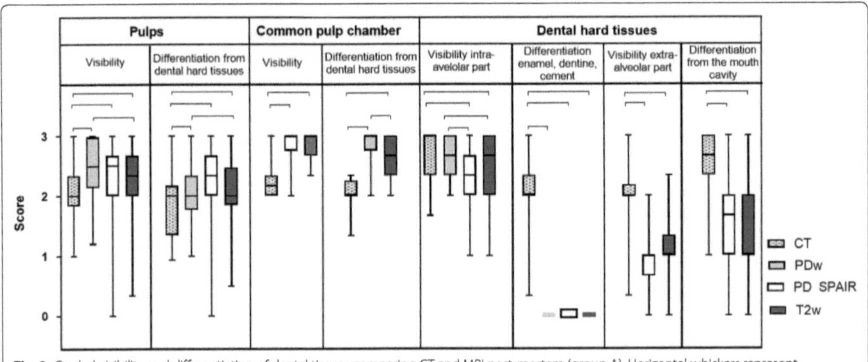

Fig. 2 Graded visibility and differentiation of dental tissues comparing CT and MRI post-mortem (group A). Horizontal whiskers represent statistically significant differences between scores. Boxes represent the interquartile range and vertical whiskers the range. *CT* computed tomography, *PDw* proton-density weighted, *PD SPAIR* proton-density weighted spectral attenuated inversion recovery, *T2w* T2 weighted

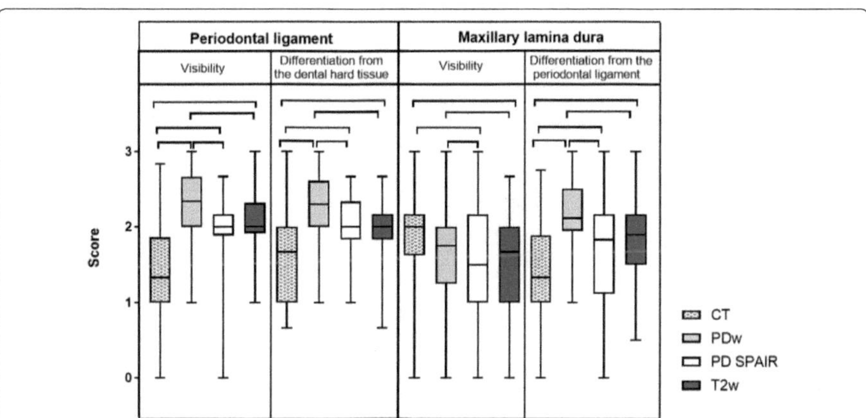

Fig. 3 Graded visibility and differentiation of periodontal tissues comparing CT and MRI post-mortem (group A). Horizontal whiskers represent statistically significant differences between scores. Boxes represent the interquartile range and vertical whiskers the range. *CT* computed tomography, *PDw* proton-density weighted, *PD SPAIR* proton-density weighted spectral attenuated inversion recovery, *T2w* T2 weighted

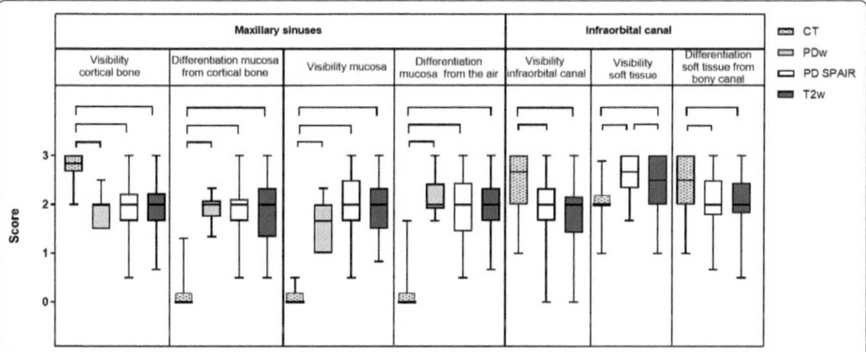

Fig. 4 Graded visibility and differentiation of adjacent tissues comparing CT and MRI post-mortem (group A). Horizontal whiskers represent statistically significant differences between scores. Boxes represent the interquartile range and vertical whiskers the range. *CT* computed tomography, *PDw* proton-density weighted, *PD SPAIR* proton-density weighted spectral attenuated inversion recovery, *T2w* T2 weighted

and was not significantly different among all imaging techniques.

Regarding the structures' visibility and differentiation from surrounding tissues, CT proved to be the superior imaging modality to display dental hard (enamel, cementum and dentine) and bony tissues (maxillary bone, infra-orbital canal): regarding dental structures, the visibility of dental hard tissues inside the maxillary bone (intra-alveolar part of the tooth), the differentiation of all dental hard tissues themselves, the visibility of the dental clinical crown and the delineation against the mouth cavity were rated higher in CT ($P \leq 0.001$) compared to all MRI sequences (Fig. 2). Differentiation of the dental hard tissues and the oral cavity was only visible in MRI when hyperintense saliva or the tongue were next to the hypointense cheek teeth displayed,

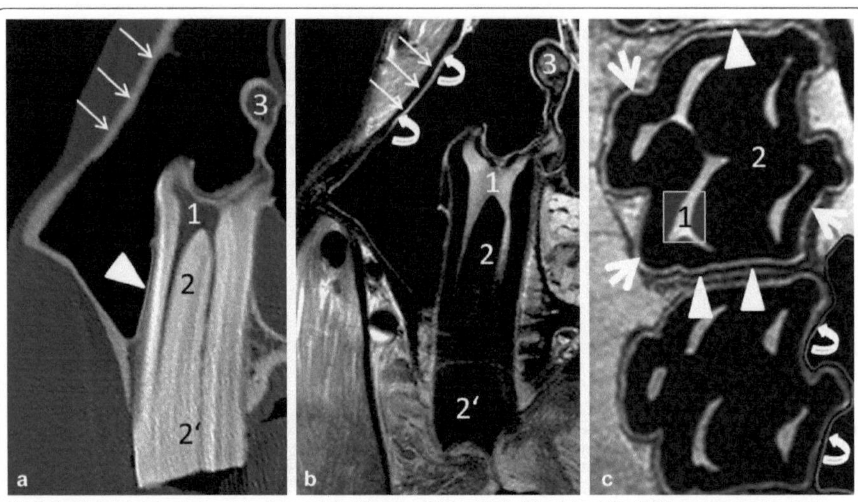

Fig. 5 Dental, periodontal and adjacent structures evaluated. Transversal CT scan (**a**), T2w (**b**) and dorsal PDw (**c**) MRI sequences post-mortem (group A). 1 = pulp; 2 = intra-alveolar part of the dental hard tissues; 2′ = extra-alveolar part of the dental hard tissues; 3 = soft tissue inside the infra-orbital canal; thick arrows = periodontal ligament; arrowheads = cortical alveolar bone; thin arrows = cortical maxillary bone; curved arrows = sinus' mucosa

resulting in low median scores for PD SPAIR and T2w sequences (Fig. 2). Figures 3 and 4 show excellent visibility of the maxillary lamina dura, cortical bone and the infra-orbital canal on CT images.

In contrast to the CT images, 3.0 T MRI was the better imaging technique to display soft tissues. The visibility and delineation of pulp, the common pulp chamber, the PDL, the mucosa of the sinuses and the infra-orbital canal's soft tissue achieved significantly better scores in MRI ($P \leq 0.0001$) than CT. Nevertheless, the delineation of soft tissues against bony structures (e.g. the infra-orbital canal and the cortical bone of the sinuses) was only visible due to the hyperintense mucosa overlaying the hypointense outlining of the bone in MRI images. When comparing the MR sequences, the differentiation of dental soft tissues (pulps, common pulp chamber, PDL) from adjacent tissues was superior in PD SPAIR and significantly improved in PDw ($P \leq 0.001$) sequences compared to T2w images (Figs. 2 and 3). Significant differences between the PDw and PD SPAIR sequence scores were evident for the periodontal apparatus: both the visibility ($P < 0.0001$) and the differentiation of the PDL from the dental hard tissues ($P < 0.0001$) and the maxillary lamina ($P < 0.001$) was significantly higher in PDw than PD SPAIR images (Figs. 3 and 4).

Comparison of the image quality and defined visibility of the structures in vivo, post-mortem and frozen-thawed (group A, B and C)

Scores for the pulp, the PDL, the mucosa of the maxillary sinuses and the soft tissue of the infra-orbital canal were compared among group A, B and C and within the CT and the MRI. The PDw sequences of group B were compared with those of group A and C to compare the MRI scores among the different specimens' conditions. The same applied to the PD SPAIR and T2w sequences.

Image quality scores

All the CT and MR images evaluated revealed good quality scores of > 2, including heads that had been frozen-thawed. Nevertheless, the image quality parameters differed between horses alive, post-mortem and frozen-thawed: image sharpness was rated significantly higher in CT ($P \leq 0.001$, median score for group B = 2.32, group B = 2.66) and MRI ($P \leq 0.005$, median score for group B = 2.13, group A = 2.33) in horses that were examined directly post-mortem than in live horses. Group C revealed median scores of 2.41 for CT and 2.24 for MRI without significant differences to the CT or MRI scores of group B (P=0.16, P=0.31) and A (P=0.11, P=0.23). Scores assessed for image noise did not differ significantly in CT imaging (P (A vs. B)=0.53; P (A vs. C)=0.40; P (B vs. C)=0.28) or in MRI (P (A vs. B)=0.37; P (A vs. C)=0.21; P (B vs. C)=0.30). Image contrast showed the best image quality scores, with values over 2.5 in CT (median score of 2.78 in group B, 2.72 in A and 2.65 in C) without significant differences among the groups (P (A vs. B)=0.56; P (A vs. C)=0.22; P (B vs. C)=0.54). High-field MRI showed very good score values for image contrast in group B (median score of 2.8) and A (median score of 2.75). Both groups displayed superior image contrast scores compared to group C (median score of 2.61) but these differences were not significant (P (B vs. C)=0.33; P (A vs. C)=0.39).

Structures' visibility scores

The CT scores for the visibility of pulp (P (A vs. B)=0.12; P (A vs. C)=0.46; P (B vs. C)=0.79) and the soft tissue inside the infra-orbital canal (P (A vs. B)=0.07; P (A vs. C)=0.15; P (B vs. C)=0.67) showed good score values in all groups without significant differences among the different groups and the MRI scores (Fig. 6) for the visibility of pulp (P (A vs. B)=0.67; P (A vs. C)=0.07; P (B vs. C)=0.08) and the infra-orbital canal's soft tissue (P (A vs. B)=0.59; P (A vs. C)=0.08; P (B vs. C)=0.30). When comparing scores for the PDL, the CT scores did not differ significantly either (P (A vs. B)=0.06; P (A vs. C)=0.19; P (B vs. C)=0.32). By contrast, the MRI showed significantly higher PDL score values in group B compared to group A (P=0.006) or C (P=0.001). Whereas the mucosa of the sinuses was not evident in CT scans of group A and B, some CT slices of heads frozen-thawed allowed the visualization of mucosa, resulting in higher score values. Nevertheless, the visibility scores of the mucosa were not significantly different between the groups in the CT scans. Regarding MRI, the mucosa had the best visualization in live horses (group B, Fig. 7a), significantly higher compared to group C ($P \leq 0.001$).

Inter-observer reliability

Calculation of inter-rater agreement for all raters showed good agreement in CT, PDw and PD SPAIR ($P<0.0001$, Table 4), with a Kappa between 0.69 and 0.71. The inter-rater agreement for T2w was moderate between all raters with a Kappa of 0.59. Whereas the agreement between rater 2 and 3 and between observer 1 and 3 (despite T2w with good agreement) was very good in all imaging techniques, results for rater 1 and 2 just achieved good agreement. Rater 3 showed a tendency to evaluate all MRI sequences and CT scans with higher scores than the other two observers. Further evaluation of those tendencies showed that they did not affect the significance of the inter-modality comparison.

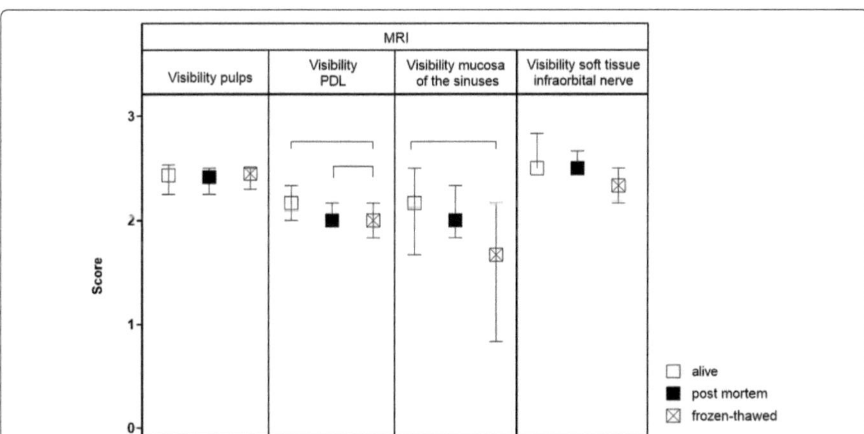

Fig. 6 Graded MRI visibility in horses alive (group B), post-mortem (group A) and frozen-thawed (group C). Horizontal whiskers show statistically significant differences between scores. Boxes represent the interquartile range and vertical whiskers the range. *MRI* magnetic resonance imaging, *PDL* periodontal ligament

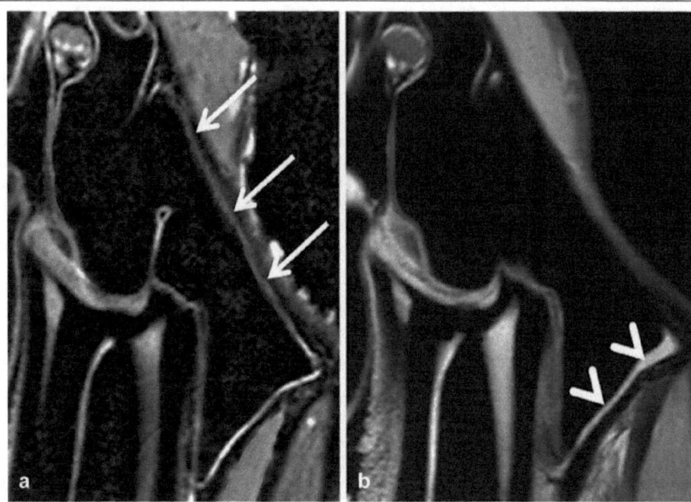

Fig. 7 MRI findings in the same horse alive (**a**, group B) and post-mortem (**b**, group A). Both images display transversal T2w scans of a 209 (cheek tooth) and the periodontal structures. Arrows show the hyperintense sinus mucosa (**a**). Image noise is visible as hyperintense, cloudy signal within the hypointense sinus and the dental hard tissues (**a**) due to small patient's movements. Arrowheads reveal thickened hyperintense mucosa (**b**) in the horse post-mortem

Table 4 Inter-rater agreement with weighted Kappa (wk) for each modality (P < 0.0001)

Modality		Raters compared			All raters
		Rater 1/2	Rater 1/3	Rater 2/3	
CT	wk	0.7750	0.8517	0.8695	0.7043
PDw	wk	0.7330	0.8135	0.8606	0.6921
PD SPAIR	wk	0.7495	0.8297	0.8841	0.7101
T2w	wk	0.6833	0.7921	0.8376	0.5953

CT computed tomography, *PDw* proton-density weighted, *PD SPAIR* proton-density weighted spectral attenuated inversion recovery, *T2w* T2 weighted

Discussion

Methodology

Recent studies described the qualities of CT [3, 9] or MRI [2, 20, 21] for the diagnosis of equine dental pathologies without any objective grading system to evaluate the different structures. The four-point rating scale used for the present examinations was designed to objectify results for a more detailed and less subjective comparison of CT and MRI. Whereas [18] focused on the CT and 3.0 T MRI's ability to portray paranasal sinuses, to the best of the authors' knowledge, the current study is the first evaluation which scores the quality of CT and MR images of equine dental tissues based on a grading system and compares the influence of the condition of the specimens on image quality.

Imaging techniques/settings

Due to the technical progress, the CT examinations on standing, sedated horses provide a feasible alternative to CT under general anaesthesia [22, 23] used in the current study. The anaesthetic risk is reduced in sedated patients in comparison to general anaesthesia [24]. Nevertheless, scanning horses under sedation is not a simple procedure and requires a team of veterinary support staff. The main disadvantage of the standing procedure is movement blur, which may degrade the image quality or necessitate repetition of a scan. In the current study, authors tried to minimize the time of the scanning procedure to get the best image quality, but this might also have been achieved in images of standing CTs through repeated scans. In comparison to standing CT, where helical scanning is the only availability, axial scans with a longer scanning duration could be acquired under general anaesthesia and provide increased image quality. However, no standing CT was available in the clinic where the imaging acquisitions were performed. The total time of general anaesthesia could be shortened by using standing CT in further studies which combine CT and MRI examinations. This would result in a reduced anaesthetic risk in live horses and could permit longer MRI scan times if more image planes or MRI sequences are needed.

Comparing the examination times in all groups, the time needed differed markedly between CT and MRI: the CT was 13 times faster than the time taken for all the MRI scans. Whereas the MPR in CT scans provides the opportunity to create images in every alignment after the examination, a scan for each alignment is required in the MRI, resulting in long acquisition times. Finally, longer MRI examination times were chosen to produce high-quality images. Under clinical conditions, the number of MRI alignments or resolution might be decreased to reduce scan time and keep the anaesthetic time and risk as minimal as possible [25]. Three-dimensional T1w MRI scans offer the exception to produce MPR series. Nevertheless, T1w series were not evaluated further in the current study due to decreased image quality. High-field MRI requires long T1w scans, as the T1 relaxation time is prolonged [26]. In the present study, T1w image scans might have been too short to achieve satisfactory image quality and tissue visibility scores, therefore, the T1w three-dimensional sequence was excluded.

As a principal finding, all other MRI scans acquired in the current study proved to be capable of illustrating the regions of interest. Comparing different field strengths in MRI, the 3 T, caused by a twice as high signal-to-noise ratio compared to 1.5 T, allows for improved image quality and spatial resolution within the same examination time [27, 28].

Cheek teeth selected

The maxillary 08s, 09s and 10s are the cheek teeth that show clinical signs, such as apical infections and infundibular caries, most frequently [1, 29]. As reported, the T2w, PD SPAIR and PDw scans evaluated did not acquire images in alignment with each tooth. Cheek teeth do not have the same alignments within one skull [30], so that the last upper teeth are not depicted mainly in perfect alignment [11]. Thus, selection was made for two adjacent cheek teeth with more similar angulations than the more caudally positioned upper cheek teeth to avoid low visibility and differentiation scores due only to the wrong alignment.

Scores for CT, T2w, PDw and PD SPAIR sequences in group A (post-mortem)

Comparing the image quality in all CT and MRI scans evaluated of group A, the scores showed higher noise and less sharpness in the MR images. Reasons could be found in the MR coil positioning: whereas the whole head was scanned in the CT, examination coils were placed around the region of interest in the MRI, allowing a field of view of about 25 × 25 cm. The field of view in the current study ranged from 16 to 25 cm, therefore, mispositioning might lead to a decrease of signal intensity and image quality [18].

The authors of the current study agree with other investigations [11, 15, 31] that MRI is an ideal non-invasive technique to display soft tissue structures due to the increased water content of the latter. Consequently, MRI provided good to excellent visibility and differentiation scores for the soft and periodontal dental tissue detail and contrast, such as the pulp, PDL, mucosa of the sinuses, and infra-orbital nerve and vessels. A benefit of this ability to depict delicate soft tissues such as the infra-orbital nerve and its content, in clinical cases is that previously undetected pathologies may be visualized [31] even before they become visible with osseous changes in the CT.

As PDw and PD SPAIR sequences highlight tissues with a high proton density, superiority was found for both sequences compared to the T2w scans. Thin structures, such as the PDL, that is part of the periodontal apparatus, proved to be better visualized in PDw scans than PD SPAIR sequences. Tissues such as the PDL, that have a high free proton density, show a large transverse component of magnetization, depicted in a high signal [32]. In contrast to MRI, CT reached the lowest visibility score for the PDL of all structures that were displayed in the CT. Thus, MRI (especially PDw sequences) might be the more suitable imaging technique to prove whether the PDL is still vital. This could be used for presurgical planning in cases of endodontic procedures [12] or replantations [13, 14] in apically infected cheek teeth, as neither procedure is advised in cheek teeth with avital PDLs. Further investigations to evaluate the visibility of diseased PDLs in MRI are required.

Whereas CT scores good to excellent results, differentiation of the junctions and intra-oral air content of the dental hard tissues was poor in MRI, which is in line with the results in human medicine [19]. Unsatisfactory scores of bony and dental hard tissue structures are caused by the inability of conventional MR measuring methods to compensate for the very short relaxation times in hard tissues [33]. The MRI only provides an indirect depiction of structures with low proton densities: good visibility of the hypointense maxillary cortical bone and the infra-orbital canal were only possible because of their delineation against the hyperintense mucosa of the sinuses, and visibility of the extra-alveolar part of the dental hard tissues through the delineation against hyper- and isointense tongue tissue and saliva. These results suggest that

CT is still the imaging technique of choice if osseous or dental structures are involved.

Comparisons of scores between euthanized horses (A), live horses (B) and frozen-thawed cadaver heads (C)

Similar equine research studies described severe changes in MR image quality of soft tissues after freezing [16]. To prove, whether the image quality suffered in post-mortem or frozen-thawed soft tissues in the current study, the image quality and visibility of the pulp, PDL, mucosa of the sinuses and infra-orbital canal's soft tissue were additionally evaluated in group B and C.

The results of the present study suggest that CT and MRI are excellent tools for good to excellent image qualities in all groups without significant differences in image noise and contrast. The image contrast was also satisfactory in heads that were frozen-thawed. The reasons could be that MRI does not measure signals for frozen materials in which atoms have lost mobility but yield a signal after the tissues thaw and reacquire molecular mobility [34]. These findings are in line with a previous study [16], where frozen limbs were defrosted and rescanned multiple times, resulting in no differences in the image quality of the scans. The high-resolution MRI presented and the CT examinations were susceptible to artefacts, resulting in worse image sharpness scores in group B than group A and C: small movements, through breathing and heartbeat, as was present in live horses, appeared as motion artefacts. The aim, therefore, should be to position and fixate the patient's head properly and reduce the total measurement time in horses under general anaesthesia. Heads were fixed at the table in all groups in the current study; nevertheless, slight movement in live horses could not be prevented (Fig. 7).

The CT, depicting the hard dental and bony tissues, did not differ in any score of the soft tissues that were evaluated between group A, B and C, except for the mucosa of the sinus: mucosal oedema occurred during the freezing process, resulting in thickened mucosa that was visible in the CT scans in single heads (Fig. 8). Finally, regarding MRI research on paranasal sinuses, it must be considered that mucosa could appear pathological after freezing,

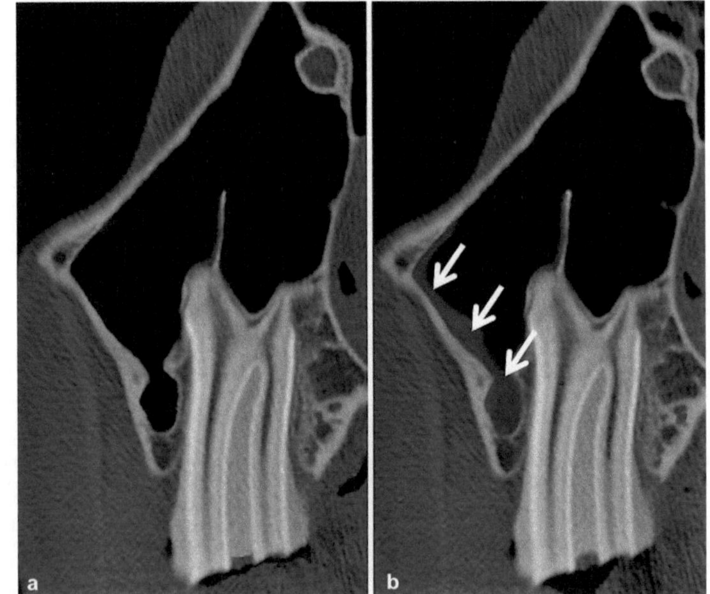

Fig. 8 CT findings in the same horse alive (**a**, group B) and frozen-thawed (**b**, group C). Both images display transversal CT scans of a 109 (cheek tooth) and the periodontal structures. Arrows show the hyperdense thickened sinus mucosa in the head frozen-thawed (**b**)

although the horse had no thickened mucosa in vivo. This might lead to false positive results. A recent MRI study of frozen human vertebral columns described that freezing and thawing leads to a decrease of signal intensity for reasons such as incomplete core specimen thawing [35]. As the core temperature was checked for the specimens in group C, these artefacts should be prevented in the current study. It has been hypothesized that autolysis and water loss may contribute to changed, more hypointense MR signals after the freezing process [16]. These findings may explain the significant decrease in MRI visibility of the PDL and mucosa of the sinuses in group C. Other processes discussed with an MR signal decrease in tissues that had been frozen-thawed, such as meat, are protein denaturation and aggregation [36]. These mechanisms were accompanied by a reduction in T1 [36] and T2 [37] values. Although quantitative MRI of meat has shown that increasing the duration of the freezing period from 2 weeks to a month at − 18 °C does not significantly enhance protein denaturation [38], slight MR signal changes were visible in the current study. The significant superiority in visibility of the PDL in live horses compared to those post-mortem (group A) may be explained by the unavailable blood flow in euthanized horses.

In contrast to the PDL and the mucosa of the sinuses, no significant score differences were visible for the infraorbital canal's soft tissue and the pulp in the MRI scans. Both structures were surrounded by hard dental and bony tissue. Even if the cells' integrity was destroyed in these structures, leading to a lower proton density, fluid cannot distribute in the oral cavity or the sinus. Thus, the surrounding structures may prevent a loss of MRI signal, depicting the extracellular fluid with a hyperintense signal.

Inter-rater agreement was good between all raters in the current examination and in a comparative study of MRI and CT regarding the equine fetlock joint [27].

Clinical relevance

Scores of group A (n = 9) allowed for a comparison of CT, T2w, PDw and PD SPAIR sequences. Whereas CT highlighted mainly the dental hard tissues and bony structures, MRI provided a perfect depiction of soft tissues, especially in PD SPAIR and PDw sequences. In clinical cases, this knowledge might help to decide on an imaging technique or a specific MRI sequence in patients with dental disorders (e.g. PDw sequences to portray the PDL). With its good to excellent scores for bony and hard dental tissues, CT remains an auspicious method to portray pathological dental progresses concerning the alveolar bone (e.g. in cases of alveolitis) and structural abnormalities of the hard dental tissues (e.g. in cases of infundibular caries). Score results show that MRI might be a promising method to evaluate the vitality of the pulp and PDL in cases of endodontic treatment and replantation of infected cheek teeth. To the best of the authors' knowledge, the correlation between MR signal intensity and the vitality of the PDL has not yet been verified in equine dentistry. Further studies with comparisons of the MR signal intensity and histological findings in diseased teeth are required to interpret the MR depiction of the PDL and pulp. In the end, the inherent problem that MRI examinations take up a long time if more sequences and orientations are necessary to evaluate pathologic processes must be considered.

Limitations

Study limitations include that both live and euthanized horses, and the cadaver heads were removed from the CT and MRI gantry between imaging sections. The reasons for the changed signals might be due to a different placement of the coils and the head relating to the isocentre of the magnetic field. Moreover, a selection bias caused by the small size of the current study population cannot be ruled out. Image interpretation might not comply with the entire population of horses.

Conclusion

The results of this experimental study suggest that CT is still the imaging technique of choice to portray bony structures and dental hard tissues. On the contrary, MRI provided a perfect depiction of soft tissues such as mucosa, the PDL and the pulpar tissue, especially in PD SPAIR and PDw sequences. The comparisons of the image quality between live, post-mortem and frozen-thawed specimens showed that image quality parameters did not suffer post-mortem or by freezing and thawing; the image sharpness was even better in these groups than in live horses and visibility scores were satisfactory for soft tissues in all specimen conditions. However, the authors' hypothesis can be confirmed: significant superiority to portray the mucosa of the sinuses and the PDL was present in live horses. As such, the current study could serve as a reference for further research investigations to decide on the best specimen condition if a specific dental or periodontal structure is to be portrayed. In this context, recent MRI studies and results of horses with cheek teeth or sinus pathologies that were acquired after freezing must be regarded critically.

Supplementary information

Supplementary information accompanies this paper at https://doi.org/10.1186/s13028-019-0495-8.

Additional file 1. Imaging techniques and parameters.

Abbreviations

CT: computed tomography; e.g.: exempli gratia = for example; MPR: multiplanar reconstruction; MR: magnetic resonance; MRI: magnetic resonance imaging; PDL: periodontal ligament; PD SPAIR: proton density weighted spectral attenuated inversion recovery; PDw: proton density weighted; T1w: t1 weighted; T2w: t2 weighted; TE: echo time; TR: repetition time; WL: window level; WW: window width.

Acknowledgements

The authors thank Dr. Karl Rohn, Department of Biometry, Epidemiology and Information Processing, University of Veterinary Medicine, Hannover, Germany for assistance with statistical analyses.

Prior publication

An abstract containing parts of the material included was presented at the 26th European Veterinary Dental Forum in Malaga, Spain, 18–20 May 2017.

Authors' contributions

All authors designed the study. CR designed the four-point rating scale and wrote the manuscript. CR and MH both optimized CT settings and acquired CT and MR data. MH supported and supervised the CT and MR data acquisition. ABZ advised CR in dental nomenclature. ABZ and MH contributed to data and image analysis and interpretation. ABZ, MH and BO contributed preparing the manuscript. Statistical analyses were acquired by CR (veterinarian) and KR, a veterinary statistician who additionally proofed the evaluations CR performed. All authors read and approved the final manuscript.

Funding

The authors received no financial support for the research, authorship, and/or publication of this paper.

Availability of data and materials

The datasets used and/or analysed during the current study are available from the corresponding author on reasonable request.

Ethics approval and consent to participate

Studies were approved by the Lower Saxony State Office for Consumer Protection and Food Safety in accordance with the German Animal Welfare Law (LAVES– Reference number: 33.12-42502.04.14/1644).

Consent for publication

Not applicable.

Competing interests

The authors declare that they have no competing interests.

Received: 19 May 2019 Accepted: 2 December 2019
Published online: 10 December 2019

References

1. Bühler M, Fürst A, Lewis FI, Kummer M, Ohlert S. Computed tomographic features of apical infection of equine maxillary cheek teeth: a retrospective study of 49 horses. Equine Vet J. 2014;46:468–73.
2. Gerlach K, Ludewig E, Brehm W, Gerhards H, Delling U. Magnetic resonance imaging of pulp in normal and diseased equine cheek teeth. Vet Radiol Ultrasound. 2013;54:48–53.
3. Veraa S, Voorhout G, Klein WR. Computed tomography of the upper cheek teeth in horses with infundibular changes and apical infection. Equine Vet J. 2009;41:872–6.
4. Casey MB, Pearson GR, Perkins JD, Tremaine WH. Gross, computed tomographic and histological findings in mandibular cheek teeth extracted from horses with clinical signs of pulpitis due to apical infection. Equine Vet J. 2015;47:557–67.
5. Tremaine WH, Dixon PM. A long-term study of 277 cases of equine sinonasal disease. Part 1: details of horses, historical, clinical and ancillary diagnostic findings. Equine Vet J. 2001;33:274–82.
6. Manso-Díaz G, García-López JM, Maranda L, Taeymans O. The role of head computed tomography in equine practice. Equine Vet Educ. 2015;27:136–45.
7. Kraft SL, Gavin P. Physical principles and technical considerations for equine computed tomography and magnetic resonance imaging. Vet Clin N Am Equine Pract. 2001;17:115–30.
8. Henninger W, Mayrhofer E. 3D-reconstruction of CT-images in horses. Vet Radiol Ultrasound. 1999;40:184.
9. Henninger W, Frame EM, Willmann M, Simhofer H, Malleczek D, Kneissl SM. CT features of alveolitis and sinusitis in horses. Vet Radiol Ultrasound. 2003;44:269–76.
10. Schoppe C, Hellige M, Rohn K, Ohnesorge B, Bienert-Zeit A. Comparison of computed tomography and high-field (3.0 T) magnetic resonance imaging of age-related variances in selected equine maxillary cheek teeth and adjacent tissues. BMC Vet Res. 2017;13:280.
11. Gerlach K, Brehm W, Gerhards W, Ludewig E. Diagnostik von Erkrankungen der Backenzähne des Pferdes mittels Magnetresonanztomographie. Pferdeheilkunde. 2011;27:711–8.
12. Lundström T, Wattle O. Description of a technique for orthograde endodontic treatment of equine cheek teeth with apical infections. Equine Vet Educ. 2016;28:641–52.
13. Staszyk C. Extraoral endodontic treatment and replantation of equine cheek teeth: anatomical, histological and molecular biological considerations. In: Proceedings of the 26th European veterinary dental forum. Malaga, Spain. 2017.
14. Stoll M, Pearce C. Extraoral endodontal treatment and replantation of equine cheek teeth. Part 2: technique and materials. In: Proceedings of the 26th European veterinary dental forum. Malaga, Spain. 2017.
15. Gaudino C, Cosgarea R, Heiland S, Csernus R, Beomonte Zobel B, Pham M, et al. MR-imaging of teeth and periodontal apparatus: an experimental study comparing high-resolution MRI with MDCT and CBCT. Eur Radiol. 2011;21:2575–83.
16. Bolen GE, Haye D, Dondelinger RF, Massart L, Busoni V. Impact of successive freezing-thawing cycles on 3-T magnetic resonance images of the digits of isolated equine limbs. AJVR. 2011;72:780–90.
17. Floyd M. The modified Triadan system: nomenclature for veterinary dentistry. J Vet Dent. 1991;8:18–9.
18. Kaminsky J, Bienert-Zeit A, Hellige M, Rohn K, Ohnesorge B. Comparison of image quality and in vivo appearance of the normal equine nasal cavities and paranasal sinuses in computed tomography and high field (3.0T) magnetic resonance imaging. BMC Vet Res. 2016;12:13.
19. Assaf AT, Zrnc TA, Remus CC, Schönfeld M, Habermann CR, Riecke B, et al. Evaluation of four different optimized magnetic-resonance-imaging sequences for visualization of dental and maxillo-mandibular structures at 3 T. J Craniomaxillofac Surg. 2014;42:1356–63.
20. Arencibia A, Vazquez JM, Jaber R, Gil F, Ramirez JA, Rivero M. Magnetic resonance imaging and cross-sectional anatomy of the normal equine sinuses and nasal passages. Vet Radiol Ultrasound. 2000;41:313–9.
21. Illenberger N, Brehm W, Ludewig E, Gerlach K. Presentation of age-related changes of the pulp horns in equine upper cheek teeth by MRI. Pferdeheilkunde. 2013;29:183–8.
22. Dakin SG, Lam R, Rees E, Mumby C, West C, Weller R. Technical set-up and radiation exposure for standing computed tomography of the equine head. Equine Vet Educ. 2014;26:208–15.
23. Strauch S, Schwarzer J, Kau S. Computed tomography in equine dentistry—a survey. Pferdespiegel. 2019;22:23–36.
24. Freeman SL, England GC. Investigation of romifidine and detomidine for the clinical sedation of horses. Vet Rec. 2000;147:507–11.
25. Weishaupt D, Köchli VD, Marincek B. How does MRI work?. 6th ed. Heidelberg: Springer Medizin Verlag; 2009.
26. Merkle EM, Dale BM. Abdominal MRI at 3.0 T: the basics revisited. AJR Am J Roentgenol. 2006;186:1524–32.
27. Hontoir F, Nisolle JF, Meurisse H, Simon V, Tallier M, Vanderstricht R, et al. A comparison of 3-T magnetic resonance imaging and computed tomography arthrography to identify structural cartilage defects of the fetlock joint in the horse. Vet J. 2014;199:115–22.
28. Saupe N, Prüssmann KP, Luechinger R, Bösiger P, Marincek B, Weishaupt D. MR imaging of the wrist: comparison between 1.5- and 3-T MR imaging—preliminary experience. Radiology. 2005;234:256–64.
29. Dacre I, Kempson S, Dixon PM. Equine idiopathic cheek teeth fractures. Part 1: pathological studies on 35 fractured cheek teeth. Equine Vet J. 2007;39:310–8.

30. Dixon PM, Du Toit N. Dental anatomy. In: Easley J, Dixon PM, Schumacher J, editors. Equine dentistry. 3rd ed. Philadelphia: WB Saunders; 2011. p. 25–48.
31. Kaminsky J, Bienert-Zeit A, Hellige M, Ohnesorge B. 3 Tesla magnetic resonance imaging of the equine nasal cavities, paranasal sinuses and adjacent anatomical structures in 13 healthy horses. Pferdeheilkunde. 2014;29:183–8.
32. Tucker RL, Sampson SN. Magnetic resonance imaging protocols for the horse. Clin Tech Equine Pract. 2007;6:2–15.
33. Detterbeck A, Hofmeister M, Hofmann E, Haddad D, Weber D, Hölzing A, et al. MRI vs. CT for orthodontic applications: comparison of two MRI protocols and three CT (multislice, cone-beam, industrial) technologies. J Orofac Orthop. 2016;77:251–61.
34. Widmer WR, Buckwalter KA, Hill MA, Fessler JF, Ivancevich S. A technique for magnetic resonance imaging of equine cadaver specimens. Vet Radiol Ultrasound. 1999;40:10–4.
35. Kurmis AP, Slavotinek JP, Barber C, Smith L, Fazzalari NL. An unusual MR signal reduction artefact in an incompletely thawed cadaver spine specimen. Radiography. 2009;15:86–90.
36. Evans SD, Nott KP, Kshirsagar AA, Hall LD. The effect of freezing and thawing on the magnetic resonance imaging parameters of water in beef, lamb and pork meat. Int J Food Sci Technol. 1998;33:317–28.
37. Renou JP, Foucat L, Bonny JM. Magnetic resonance imaging studies of water interactions in meat. Food Chem. 2003;82:35–9.
38. Guiheneuf TM, Parker AD, Tessier JJ, Hall LD. Authentication of the effect of freezing/thawing of pork by quantitative magnetic resonance imaging. Magn Reson Chem. 1997;35:112–8.

Publisher's Note

Springer Nature remains neutral with regard to jurisdictional claims in published maps and institutional affiliations.

4 Übergreifende Diskussion

4.1 Untersuchungsmaterial

Für die Untersuchung der normal-anatomischen Verhältnisse und der Darstellbarkeit ausgewählter maxillärer Prämolaren und Molaren sowie ihrer benachbarten Strukturen wurden zwölf Warmblutpferde unterschiedlichen Alters und Geschlechts ausgewählt. Um altersvergleichende Unterschiede der Probanden herauszuarbeiten, wurden die Pferde in drei Altersgruppen eingeteilt: "jung" (2-5 Jahre), "mittleres Alter" (6-15 Jahre) und "alt bis senil" (≥ 16 Jahre), jeweils mit n = 4. Alle Pferde waren zum Zeitpunkt der klinischen Untersuchungen allgemeingesund und zeigten klinisch keine Anzeichen einer dentogen bedingten Erkrankung. Die Euthanasie der Probanden erfolgte innerhalb einer anderen wissenschaftlichen Studie in Allgemeinanästhesie. Nicht alle der ursprünglich für die Untersuchungen eingeplanten Pferde, konnten aus unterschiedlichen Gründen abschließend für die Studie genutzt werden: da im CT oder MRT in vivo ein hochgradiger Flüssigkeitsspiegel im rechten Sinus maxillaris sichtbar war (n = 1), der Kopf post mortem nicht mehr zur Verfügung stand (n = 1), oder mehrere Pulpen der zu untersuchenden Backenzähne im CT in Form von hypodensen Gaseinschlüssen eine Pulpitis zeigten (n = 1).

Während neun Pferdeköpfe post mortem der CT- und MRT-Untersuchung unterzogen wurden, sind Aufnahmen in vivo und Untersuchungen an aufgetauten Köpfen jeweils an drei der neun Pferdeschädeln durchgeführt worden. Trotz der kleinen Studienpopulation wurden – aufgrund der Vielzahl der Schnitte, den verschiedenen Schnittbildebenen (transversal und dorsal) und den unterschiedlichen MRT-Sequenzen – insgesamt 14040 Parameter (8424 für Pferde direkt post mortem, jeweils 2808 für Präparate in vivo und eingefrorene-wieder-aufgetaute Köpfe) ausgewertet. Hierdurch war ein Vergleich der CT- und MRT-Bilder mit Hilfe eines Auswertungsschlüssels für die Studienpopulation möglich. Um das Score-System auf eine bestmögliche Bildqualität anzuwenden, wurden im Vorversuch an Kadaverköpfen die Geräteeinstellungen für die Darstellung von Zähnen und angrenzendem Gewebe sowohl für die CT als auch die MRT optimiert.

4.2 Auswahl der Oberkieferbackenzähne

Für die Benennung der Zähne wurde das modifizierte Triadan-System genutzt (FLOYD 1991). Jeweils der 108er, 208er, 109er und 209er wurde bei jedem der neun Probanden mittels CT- und MRT-Untersuchungen befundet (n = 36). Die in dieser Studie angefertigten CT- und MRT-Bilder zur physiologischen Darstellung der Backenzähne und angrenzender Strukturen können bei erkrankten Pferden als normal-anatomischer Vergleich genutzt werden. Die 08er und 09er gehören zu den Oberkieferbackenzähnen, die am häufigsten Pathologien wie Infundibularkaries (HENNINGER et al. 2003), apikale Infektionen oder Zahnfrakturen (BAKER 1999, BÜHLER et al. 2014) aufweisen. In der Literatur wird als Grund für die gehäuften Pathologien der 08er und 09er die Position der Zähne angeführt: DACRE et al. (2007) argumentierten, dass Backenzähne, die in der Mitte der Kauleiste liegen, einem höheren Druck beim Kauen standhalten müssen als peripher positionierte. Bei den permanenten 09ern ist zudem eine hohe mechanische Belastung besonders früh gegeben, da diese Oberkieferzähne bereits mit einem Pferdealter von etwa einem halben Jahr ohne vorherige Milchzähne durchbrechen.

Während der okklusale Anteil des maxillären 06er nach kaudal und der des 10er und 11er deutlich nach rostral gerichtet ist, zeigen die mittig liegenden 08er und 09er eine nahezu perpendikuläre Ausrichtung zum harten Gaumen (STASZYK 2015). Obwohl die Angulation von P4 und M1 nicht exakt die gleiche ist, erleichtert ihre ähnliche Winkelung die Ausrichtung der Schnittbildebene bei der Erstellung der MRT-Aufnahmen. Im Gegensatz zur MPR (CT) und den 3D-MRT-Aufnahmen war eine Anpassung der Schnittbildebene an einzelne Zähne nach Akquisition der PDw-, T2w- und fettunterdrückten MRT-Bildern in der vorliegenden Studie nicht mehr möglich. Der Vorteil bei der Verwendung des 08er und 09er liegt folglich darin, dass mit einer MRT-Sequenz zwei Zähne nahezu orthograd abgebildet werden können, da der Winkel beider Oberkieferbackenzähne fast identisch ist. Dies ist zeitsparender als für jeden Zahn einzelne MRT-Sequenzen anzufertigen.

4.3 Versuchsablauf

Mit Hilfe von Vorversuchen an drei toten Pferden konnte der Ablauf der Lagerung optimiert und somit die Dauer der Untersuchungen verkürzt werden. Diese betrug im Hauptversuch für die aufeinander folgenden CT- und MRT-Untersuchungen insgesamt zwischen 103 und 120 Minuten (Lagerung CT ± SD, 5 ± 3 min; CT 2 ± 5 min; MRT 89 ± 9 min, Lagerung zwischen CT und MRT 12 ± 3 min). Die Euthanasie der Pferde erfolgte im Vor- und Hauptversuch unabhängig von der vorliegenden Studie auf Grund von schwerwiegenden Erkrankungen des Bewegungsapparates oder des Magen-Darm-Traktes. Kein Versuchspferd zeigte klinische Symptome einer Erkrankung der Sinus oder der Zähne. Die Probanden wurden zunächst auf einem stationären CT-Tisch in rechter Seitenlage und im Anschluss in Rückenlage auf einem speziellen MRT-Tisch positioniert. Im CT erwies sich eine Lagerung auf der Seite als zeitsparender im Vergleich zu einer Lagerung der Pferde auf dem Rücken, da die Beine nicht fixiert werden mussten. Dadurch war die Lagerung bei den beiden bildgebenden Verfahren zwar nicht einheitlich, aber es konnte eine effiziente, stabile und zeitsparende Lagerung im CT erzielt werden, die auch für klinische Patienten angewandt werden kann. Für Kadaverstudien oder Untersuchungen an aufgetauten Präparaten ist die Untersuchungszeit unbedeutend. Bei Patienten oder Pferden in vivo ist eine kürzere Anästhesiezeit jedoch mit einem geringeren Narkoserisiko und einer risikoärmeren Aufstehphase verbunden (JOHNSTON et al. 2002, MERKLE u. DALE 2006, CHANG et al. 2008). Für den Fall, dass die in vivo Studien im Ergebnis bessere Bildqualitäten liefern würden als die der post mortem Präparate, wurde der Versuch von Beginn an so geplant, dass die Untersuchungen mit einer möglichst geringen Anästhesiezeit bei bestmöglicher Bildqualität umgesetzt werden konnten. Das in diesem Versuch genutzte Protokoll kann deshalb auch in nachfolgenden Forschungsarbeiten an lebenden Pferden und Patienten angewandt werden. Die mit CT- und MRT-Untersuchung sowie der Lagerung einhergehende Dauer der Allgemeinanästhesie von etwa zwei Stunden würde bei einem klinischen Fall die Möglichkeit der anschließenden chirurgischen Therapie dennoch erheblich einschränken. Bislang ist in der Pferdemedizin die CT-Untersuchung zur Abklärung von Zahnerkrankungen mit anschließender chirurgischer Intervention etabliert

(VERAA et al. 2009). Neben der sehr kurzen Untersuchungszeit ermöglicht die CT eine Abbildung des gesamten Kopfes und somit ein großes Untersuchungsfeld (FOV). Bei CT-Untersuchungen größerer Köpfe erhöht sich bei gleichbleibender Schichtdicke lediglich die Anzahl der Schnitte und somit die Untersuchungszeit. In der MRT-Untersuchung ist für die Akquisition der Bilder das Anlegen von Empfängerspulen erforderlich. Die in dieser Studie genutzten Spulen (SENSE Flex M und L ®) grenzten das FOV in der vorliegenden Studie auf eine Größe von ca. 30 cm ein. Außerhalb der Empfängerspulen nimmt die Bildqualität deutlich ab. Aus diesem Grund wurden die MRT-Spulen immer zentral über die 08er und 09er des Oberkiefers gelegt, um Fehlpositionierungen der Empfangsspule zu verhindern. Die Größe der genutzten, geschlossenen MRT-Gantry und die Konformation des mobilen MRT-Tisches ermöglichen für die Abbildung des Kopfes ausschließlich eine Lagerung des Pferdes auf dem Rücken. Somit war die Kopfpositionierung im CT und MRT zwar nicht einheitlich, die Rückenlage des Pferdes bietet jedoch auch Vorteile: Hyperintense Flüssigkeitsansammlungen können besser von Schleimhautschwellungen in den Sinus abgegrenzt werden, da sich Flüssigkeitsspiegel durch unterschiedliche Patientenlagerungen ändern. Zudem wird das Risiko einer Fazialisparese, die durch eine längere Weichteilkompression in Seitenlage entstehen kann, reduziert (WAGNER 2008). Die in der Pferdemedizin alternativ verwendeten, halboffenen Systeme mit einer Magnetfeldstärke von 0,25 Tesla ermöglichen zwar eine Abbildung des Kopfes in Seitenlage, gehen aber im Vergleich zum 3-T-Hochfeld-MRT mit einer deutlich reduzierten Bildqualität einher (BOLAS 2010). Da in der hiesigen Studie jedoch eine hohe Bildqualität für die Bewertung der einzelnen Strukturen benötigt wurde, um die verschiedenen Bildgebungsmodalitäten vergleichen zu können, war das genutzte Hochfeld-MRT besser geeignet als ein halboffenes Niederfeld-System. Abseits der Studie sind hochauflösende MRT-Bilder ebenfalls von Vorteil, da sie gegenüber qualitativ schlechteren Niederfeld-MRT-Bildern eine exaktere Befundung ermöglichen, die die weitere Behandlung und Therapie maßgeblich beeinflussen kann.

Alternativ zu der in dieser Studie durchgeführten CT-Durchführung in Allgemeinanästhesie sind durch die technischen Fortschritte auch CT-Aufnahmen am stehenden, sedierten Pferd ohne signifikante Reduzierung der Bildqualität möglich (DAKIN et al. 2014, STRAUCH et al. 2019). Im Vergleich zur Allgemeinanästhesie sind die Risiken einer Sedierung deutlich geringer (FREEMAN et al. 2000). Voraussetzung für qualitativ hochwertige Aufnahmen einer CT-Untersuchung am stehenden Patienten sind ein geeignetes Sedierungsprotokoll und erfahrene Anästhesisten: Im Gegensatz zur CT in Allgemeinanästhesie, ist bei einer CT am stehenden Pferd eine ausreichende Sedierungstiefe für die Durchführung und die Bildqualität maßgeblich, um Bewegungsartefakte durch Kopfbewegungen des Pferdes zu vermeiden. Treten Bewegungsartefakte auf, muss der CT-Scan der betroffenen anatomischen Region wiederholt werden. Dies führt zu einer Steigerung der Untersuchungszeit.

In zukünftigen Studien könnte mit Hilfe der CT-Durchführung im Stehen die Gesamtzeit der Anästhesie verkürzt werden, da eine Allgemeinanästhesie lediglich für die MRT-Aufnahmen notwendig ist. Neben einer Reduzierung des Narkoserisikos wäre so bei klinischen Patienten möglicherweise auch nach CT und MRT-Untersuchung im direkten Anschluss eine chirurgische Intervention möglich.

4.4 CT

Zunehmend wird heute - selbst in der Routinediagnostik - die CT wegen ihrer Möglichkeiten der räumlichen Darstellung für die Zahnbefundung genutzt (CASEY et al. 2015). Diese bildgebende Modalität gilt in der Human- (VANDENBERGHE et al. 2010) und Tiermedizin (HENNINGER et al. 2003) als Verfahren der Wahl, um Erkrankungen im Bereich der Zähne und angrenzender Strukturen bei unklarem Röntgenbefund zu diagnostizieren. Die CT bietet im Gegensatz zur MRT eine gute Darstellung von Zahnhartsubstanzen und knöchernen Strukturen und wurde daher in der vorliegenden Studie gezielt mit der MRT-Darstellung von Zähnen verglichen.

Für die Auswertung wurden axiale CT-Scans angefertigt. Wie von BRINKSCHULTE (2012) sowie SCHWARZ und SAUNDERS (2011) aufgezeigt, weisen die Bilder des axialen Scanmodus durch eine höhere Ortsauflösung eine bessere Qualität als die eines helikalen Scanmodus auf. Mit Hilfe einer hohen Bildmatrix von 1024 x 1024 und einer im Vergleich zu anderen Studien (VERAA et al. 2009) geringen

Schichtdicke von 1,5 mm konnten hochaufgelöste Aufnahmen erzielt werden. Die von BARBEE et al. (1987) empfohlene Schichtdicke von 1 mm bietet zwar eine höhere Kontrastauflösung, erfordert jedoch eine höhere Bildanzahl. Die Untersuchungszeit selbst wird dadurch nicht verlängert, die Zeit der Rekonstruktion am CT-Rechner erhöht sich jedoch geringfügig. Ein maßgeblicher Nachteil dünnerer Schichtdicken ist jedoch die Erhöhung des Bildrauschens (SCHWARZ u. SAUNDERS 2011), die innerhalb des in dieser Studie durchgeführten Score-Systems zu einem verschlechterten Score-Ergebnis hätte führen können. Aus diesem Grund wurde mit der Schichtdicke von 1,5 mm ein Kompromiss zwischen detailgenauen CT-Aufnahmen und möglichst geringem Bildrauschen getroffen.

Die Auswertung und der Vergleich der CT-Bilder mit den magnetresonanztomographischen Aufnahmen wurde nach transversaler und dorsaler Ausrichtung der CT-Sequenzen mittels MPR durchgeführt. So konnte eine bessere Vergleichbarkeit der Bilder zwischen den verschiedenen Bildgebungsverfahren hergestellt werden, da vergleichbare Schnittbildebenen von CT und MRT für die weitere Bewertung mittels Score-System angefertigt werden konnten.

4.5 MRT

Neben der besten Bildgebungsmodalität für die jeweils abzubildenden Strukturen, wird zugleich eine Verringerung der Strahlenexposition bei gleichbleibend guter Bildqualität gefordert (GAUDINO et al. 2011). Vor diesem Hintergrund muss die Anwendung und der Nutzen der CT zur Beurteilung dentogener und periodontaler Gewebe hinterfragt und diskutiert werden. Dies gilt besonders für den Einsatz der CT am stehend sedierten Pferd, falls durch Bewegungsartefakte CT-Aufnahmen mehrfach durchgeführt werden müssen oder eine Anwesenheit des Klinikpersonals im Untersuchungsraum notwendig ist. Die Möglichkeit auch ohne ionisierende Strahlung Bildaufnahmen anfertigen zu können, zeigt wie wichtig es ist, die Möglichkeiten der MRT für die dentale Diagnostik zu evaluieren.

Der in dieser Studie erzielte direkte Vergleich zwischen CT- und MRT-Untersuchungen veranschaulicht, welches Potential die MRT gegenüber der CT bei physiologischen Verhältnissen von Zähnen und angrenzender Gewebe bietet. Um die bestmögliche Modalität für die jeweilige Struktur zu identifizieren, wurden verschiedene MRT-Sequenzen angefertigt: T1-, T2-, protonengewichtete und fettunterdrückte Sequenzen. Diese Wichtungen bieten eine gute anatomische Übersicht dentaler und angrenzender Strukturen und wurden bereits für die Darstellung normal-anatomischer Verhältnisse des Pferdekopfes (ARENCIBIA et al. 2000, GERLACH et al. 2009), sowie der Zähne (GERLACH et al. 2013) beschrieben. In der Auswertung der Aufnahmen, wiesen die T1-gewichteten Bilder im Vergleich zu den PDw, T2w und fettunterdrückten Sequenzen häufiger ein niedrigeres Signal-Rausch-Verhältnis auf. Da das bestmögliche Bildgebungsverfahren herausgestellt werden sollte, wurden diese Aufnahmen zwar für die deskriptive Auswertung, jedoch nicht für die weitere Evaluierung mit Hilfe des Score-Systems genutzt. Die Abnahme der Bildqualität kann auf verschiedene Gründe zurückgeführt werden: Bei T1-gewichteten Spin-Echo-Sequenzen sind die Untersuchungszeiten der Hochfeld-MRTs länger als bei Magneten mit niedrigeren Feldstärken, da die T1-Relaxationszeit verlängert ist (MERKLE u. DALE 2006, WEISHAUPT et al. 2009). Um das MRT-Protokoll in möglichst kurzer Anästhesiezeit zu gestalten, wurde in der vorliegenden Studie eine 3D T1w Sequenz mit isotropen Voxeln genutzt, wodurch die Rekonstruktionsmatrix geringer ausfiel. WEISHAUPT et al. (2009) weisen darauf hin, dass eine Änderung der Sequenzeinstellungen zur Reduktion der Untersuchungszeiten häufiger mit einer geringeren Bildqualität einhergeht: wird die Schichtdicke erhöht, um die Anfertigungszeit der MRT-Bilder zu reduzieren, treten vermehrt Partialvolumeneffekte auf. Diese MRT-Artefakte gehen mit Bildkontrastverlust zwischen zwei benachbarten Geweben einher. Ursächlich hierfür sind unterschiedliche Gewebe, die im gleichen Voxel gemessen werden. Wenn Spins von Fett und Wasser das gleiche Voxel belegen, beeinflussen sich ihre Signale durch destruktive Interferenzen: Das MR-Signal für Wasser wird hierbei durch das Fettsignal im gleichen Voxel teilweise gelöscht, sodass das entstehende Voxel nur noch das Fettsignal enthält. Je größer die Schichtdicke, desto größer das Risiko, dass ein „partial-volume-effect" entsteht, da sich zwei unterschiedliche Gewebearten ein Voxel teilen (WEISHAUPT et al. 2009). Trotz der verminderten Bildqualität

wurden die T1-gewichteten Aufnahmen nach den Vorversuchen nicht aus dem Protokoll entfernt, da sie im Vergleich zu den anderen MRT-Wichtungen durch die 3D-Daten den Vorteil bieten, Schnittbildebenen durch MPR in der Ausrichtung jedes einzelnen Zahnes anfertigen zu können. So konnte sichergestellt werden, dass ein nicht durchgängig abgebildetes Ligamentum periodontale in einer der anderen Wichtungen der Bildausrichtung geschuldet war, wenn es sich in der T1-gewichteten Aufnahme mit Hilfe einer noch besser angepassten Ausrichtung durchgängig darstellte.
Hinsichtlich des zeitlichen Aufwandes sind die gewählten Einstellungen des MRT-Protokolls auch im klinischen Arbeitsablauf anwendbar. Für spezifische Fragestellungen, die möglicherweise eine anschließende chirurgische Therapie beinhalten, könnten jedoch auch einzelne Sequenzen für die Darstellung der gewünschten Struktur gewählt, und somit die Anästhesiedauer weiter reduziert werden.
Speziell in der Endodontie, einem Teilgebiet der Zahnmedizin, ist der behandelnde Tierarzt auf gute Kenntnisse des komplexen Pulpensystems angewiesen, die in der vorliegenden Studie untersucht wurden. Die erste Untersuchung, die zur equinen Endodontie der Backenzähne veröffentlicht wurde, beschreibt eine Apikektomie und retrograde endodontische Behandlung der Pulpen bei 12 Pferden (SIMHOFER et al. 2008). Die langfristigen Therapieergebnisse dieser Studie zeigten jedoch ein unbefriedigendes Resultat. LUNDSTRÖM u. WATTLE (2016) beschrieben ein endodontisches, orthogrades Verfahren, mit dem entzündete oder nekrotische Pulpen behandelt wurden. Mittels eines Bohrers wurde okklusal der Zugang zur betroffenen Pulpe geschaffen und eine endodontische Füllung vorgenommen. Neben der Anatomie der einzelnen Pulpe, ist auch der Winkel der Pulpe zur Okklusionsfläche für den Bohrwinkel von Bedeutung. Zudem muss das mögliche Vorhandensein einer gemeinsamen Pulpenhöhle bzw. der Kommunikation einzelner Pulpen untereinander berücksichtigt werden: Wird nur eine Pulpe behandelt und die Pulpen konfluieren z. B. apikal zu einer gemeinsamen Pulpenhöhle (WINDLEY et al. 2009, KOPKE et al. 2012), kann die Infektion einer Pulpenhöhle auf andere Pulpen übergreifen. In diesen Fällen sollten alle betroffenen Pulpenpositionen eröffnet und

behandelt werden (LUNDSTRÖM et al. 2012). Während LUNDSTRÖM u. WATTLE (2016) auf Röntgenverfahren und computertomographische Verfahren zurückgriffen, zeigen die Ergebnisse der vorliegenden Studie, dass die Darstellung der Pulpen und der gemeinsamen Pulpenhöhle mit Hilfe der MRT eine bessere Bildqualität lieferte. So können durch die MRT auch kleine Verbindungen zwischen den Pulpen erkannt werden. Für den langfristigen Erfolg einer endodontischen Behandlung spielt neben dem Auffinden der Pulpen auch eine Überprüfung der Vitalität des Ligamentum periodontale eine entscheidende Rolle, das mit Hilfe der MRT-Sequenzen hervorragend abgebildet werden kann (s. Manuskript I und II). Zusammenfassend könnte die MRT zur Planung chirurgischer Verfahren im Bereich der Pulpen oder des Ligamentum periodontale durch die exzellente Darstellung der Weichteilstrukturen zum Therapieerfolg beitragen.

4.6 Score-basierter Vergleich von 3-T-MRT und CT

Der Vergleich von MRT- und CT-Bildern erfolgte sowohl deskriptiv als auch mit Hilfe eines an KAMINSKY et al. (2016) angelehnten Auswertungsschlüssels. Das Score-System enthielt dabei vier verschiedene Bewertungsscores (von 0 = schlecht bis 3 = exzellent) für die jeweiligen Strukturen. In einer vergleichbaren Studie von GAUDINO et al. (2011) wurden Unterkieferzähne von Schweinen und anliegende Strukturen in ihrer Darstellbarkeit im MRT, Multidetektor-CT und Cone-Beam-CT verglichen. Die Beurteilung der Bilder erfolgte hierbei mit einer Bewertungsskala, die fünf Scorestufen beinhaltete (von 1 = exzellent bis 5 = nicht sichtbar). Um in der vorliegenden Studie zu verhindern, dass häufig mit einem „mittig liegenden" Score bewertet wird, wurde das an KAMINSKY et al. (2016) angelehnte Bewertungssystem für die Beurteilung der CT- und MRT-Sequenzen bevorzugt. So zeichnet sich eher eine Tendenz für eine gute (2 oder 3) oder schlechte Bewertung (0 oder 1) für die Darstellbarkeit der dentogenen und periodontalen Strukturen ab.

Die Auswertung erfolgte durch drei unabhängige Gutachter (Manuskript I und II). Im Gegensatz zu einer Beurteilung, die im Gruppenkonsens durchgeführt wird (VALLANCE et al. 2011), ermöglicht eine unabhängige Bewertung, dass die Gutachter sich nicht gegenseitig beeinflussen. Trotz dieser detaillierten

Auswertungsmethodik, zeigt das *inter-rater-agreement der* vorliegenden Arbeit eine beachtliche bis exzellente Übereinstimmung zwischen den einzelnen Untersuchern. Sowohl in den CT- als auch MRT-Untersuchungen wurden Artefakte beobachtet, die teilweise die Bildqualität beeinträchtigten. Sie wurden im Rahmen der Beurteilung der Schnittbilder mit Hilfe des Score-Systems durch die einzelnen Gutachter nicht gesondert ausgewertet. Die Ursache der Artefakte unterscheiden sich auf Grund der unterschiedlichen physikalischen Grundlagen von CT und MRT, sodass kein direkter Vergleich möglich ist. Möglicherweise sind bei der Auswertung aber einzelne Parameter dadurch beeinflusst worden.

4.7 Altersvergleichende Unterschiede

Die Messungen für die Untersuchung altersbedingter Unterschiede der Pulpen und anatomischen Lagebeziehungen der Strukturen untereinander (Zahn und Infraorbitalkanal) erfolgten wie bei ILLENBERGER et al. (2013) durch einen Untersucher. Die Längenmessungen wurden dreimal zu unterschiedlichen Zeitpunkten wiederholt, um Messfehler zu minimieren. Dabei waren die Längenangaben der vorherigen Untersuchungen in jedem Durchlauf verblindet. Im Anschluss wurde aus den drei Werten ein Mittelwert gebildet und mit Hilfe eines guten *intra-rater-agreements* die Genauigkeit der Messungen überprüft und gesichert.

Zur Untersuchung der altersbedingten Unterschiede wurde mit Hilfe des Eruptionszeitpunktes das Zahnalter des jeweiligen Zahnes bestimmt. Diese Kalkulation erfolgte durch Subtraktion des frühestmöglichen Zeitpunkts des Zahndurchbruches vom Alter des Probanden (DACRE et al. 2008). Für die 08er wurde ein frühestmöglicher Durchbruch im Pferdealter von 3,5 Jahren angenommen, für die 09er von sechs Monaten (NICKEL et al. 2004). Der Eruptionszeitpunkt für die 09er variiert jedoch mit einem Pferdealter von sechs bis neun Monaten stark (DIXON et al. 2011). Aufgrund der Kalkulation des Zahnalters, könnte es – bei Kenntnis des tatsächlichen Zahnalters – zu geringfügig abweichenden Ergebnissen für die statistisch erhobenen Korrelationen zwischen Zahnalter und Längenmessungen kommen. Für die exakte Bestimmung des Zahnalters wären regelmäßige

Zahnkontrollen der Probanden ab dem 6. Lebensmonat nötig. Dies ist für die vorliegende Studie nicht realisierbar gewesen.
Die Bewertung der Pulpengröße wurde anhand der dorsalen PD-gewichteten Schnittbilder vorgenommen. Da diese aber jeweils im Abstand von 4 mm angefertigt wurden (s. Manuskript I), ist mit Messungenauigkeiten zu rechnen. Es könnten sehr feine Seitenkanäle oder Verbindungen der Pulpen untereinander aufgrund dieser Schichtdicke nicht erfasst worden sein. Eine geringere Schichtdicke, wie in den T1-gewichteten MRT-Bildern, könnte diese Ungenauigkeiten minimieren. Bei dickeren Schichtdicken, wie in den PDw, T2w und fettunterdrückten Sequenzen, sind zusätzlich „Partial-Volumen-Effekte" möglich. Da die anderen Sequenzen wie die PDw-Aufnahmen im Vergleich zu den T1w Aufnahmeserien in der vorliegenden Studie durch einen höheren Bildkontrast und eine bessere Bildschärfe dennoch eine deutliche höhere Bildqualität aufwiesen, wurden Pulpenmessungen mit Hilfe der PDw anstelle der T1w Sequenz durchgeführt. Da die Artefakte von CT und MRT in den Bewertungen der Parameter nicht gesondert beurteilt wurden, kann zur Häufigkeit möglicherweise aufgetretener Partialvolumeneffekte keine Aussage getroffen werden.
Die Abstandsmessungen zwischen dem Canalis infraorbitalis und dem apikalen Anteil der Backenzähne wurden anhand der CT-Aufnahmen durchgeführt. Diese wiesen eine geringere Schichtdicke von 1,5 mm auf. Das Risiko für Messungenauigkeiten durch fehlende Abschnitte zwischen den einzelnen Bildern, sollte somit kleiner sein als bei den Messungen in den MRT-Aufnahmen.

4.8 Darstellung und Vergleich ausgewählter Strukturen mittels CT und MRT in verschiedenen Präparatzuständen (in vivo, direkt post mortem und eingefroren-aufgetaut)

Die Anfertigung der CT- und MRT-Aufnahmen für die postmortale Untersuchung der Probanden wurde innerhalb von sechs Stunden nach Euthanasie durchgeführt, um die postmortalen Artefakte so gering wie möglich zu halten.

Wie bei MORROW et al. (2000), BRINKSCHULTE (2012) und KAMINSKY et al. (2016) beschrieben, können bei CT- oder MRT-Untersuchungen, die nach der Euthanasie durchgeführt werden, Flüssigkeitsspiegel in den Sinus als postmortales Artefakt auftreten. Als Ursache wird hierfür ein erhöhter hydrostatischer Druck in der nasalen Schleimhaut in Seiten- oder Rückenlage durch Lagerung des Kopfes unterhalb der Herzbasis angegeben (LUKASIK et al. 1997). Der in der vorliegenden Studie bei einem lebenden Pferd vorliegende Flüssigkeitsspiegel im Sinus maxillaris ist dagegen als pathologischer Befund zu deuten.

Um durch Blutverlust auftretende Veränderungen der Nasenschleimhaut und die damit verbundenen Veränderungen der Darstellbarkeit im CT und MRT (KAMINSKY et al. 2016) möglichst gering zu halten, wurden in der vorliegenden Studie größere Blutgefäße im Halsbereich vor ihrer Durchtrennung kopfwärts ligiert.

Auch die Temperatur des Präparates hat einen Einfluss auf die Bildqualität der im MRT dargestellten Strukturen (Manuskript II). Während bei lebenden Pferden die Temperatur des Gewebes nahe der physiologischen Körpertemperatur lag, wiesen die Präparate post mortem niedrigere Temperaturen auf. Um zu große Temperaturdifferenzen zwischen den drei Präparatzuständen zu vermeiden, wurden die eingefrorenen Köpfe so lange aufgetaut, bis ihre Temperatur nahezu den post mortem Präparaten entsprach (s. Manuskript II). Zusätzlich können am aufgetauten Präparat Proteindenaturation, Autolyse und Wasserverlust des Weichteilgewebes eine Verschlechterung des MR-Signals bedingen (SAUPE et al. 2005, BOLEN et al. 2011, HONTOIR et al. 2014). Diese Faktoren tragen möglicherweise zu der schlechteren Darstellbarkeit von PDL und Sinusschleimhaut in den aufgetauten Präparaten gegenüber lebenden oder direkt post mortem untersuchten Pferden in der vorliegenden Studie bei.

5 Zusammenfassung

Christin Röttiger

Altersvergleichende Darstellung equiner Oberkieferbackenzähne und angrenzender Strukturen mittels 3-Tesla-MRT und CT unter Berücksichtigung des Präparatzustandes

Moderne Bildgebungsverfahren, wie die CT und MRT bieten den Vorteil, Schnittbilder ohne Überlagerung zu erstellen. Während CT-Untersuchungen am Pferdekopf zur etablierten Diagnostik gehören, werden MRT-Aufnahmen zur Befunderhebung selten angefertigt. Durch die Erstellung von Schnittbildern und den direkten Vergleich beider Verfahren, konnte eine Wissensgrundlage für die Darstellbarkeit der Zahnstrukturen und benachbarter Gewebe bei physiologischen Verhältnissen geschaffen werden.

Für die Evaluierung der besten Bildqualität wurden die Köpfe von neun klinisch gesunden Pferden innerhalb von sechs Stunden post mortem mittels CT und MRT untersucht und insgesamt 36 Oberkieferbackenzähne (108, 208, 109, 209) und angrenzende Strukturen befundet. Hierfür erfolgte eine computertomographische Untersuchung in Seitenlage und im direkten Anschluss eine magnetresonanztomographische Darstellung der ausgewählten Strukturen in Rückenlage. Mit Hilfe der MPR wurden transversale und dorsale CT-Schnittbilder erstellt. Die MRT-Untersuchung bestand aus T1- (3D), T2- (dorsal und transversal), protonengewichteten und fettunterdrückten Sequenzen (dorsal und transversal).

Für die Auswertung der Aufnahmen wurden verschiedene Schnittbilder in definierten Ebenen jedes Probanden und jeder Bildgebungsmodalität ausgewählt und miteinander verglichen. Anhand eines Auswertungsschlüssels wurde dabei die Bildqualität sowie die Darstellbarkeit der Strukturen anhand von 1080 Schnittbildern von jeweils drei unabhängigen Gutachtern bewertet.

Bei den in dieser Arbeit untersuchten bildgebenden Modalitäten hat sich die CT als das beste Verfahren zur Darstellung knöcherner Strukturen und der

Zahnhartsubstanzen erwiesen. Zur Beurteilung des Weichteilgewebes hat sich die MRT-Untersuchung bewährt. Im Vergleich der einzelnen MRT-Aufnahmen erwies sich für die Darstellung der Pulpen und des Ligamentum periodontale die protonengewichtete Sequenz als beste MRT-Bildgebungssequenz.

Neben der bestmöglichen Darstellung der einzelnen Strukturen, wurden auch altersvergleichende Unterschiede mit Hilfe der CT und MRT untersucht, da besonders im Bereich der Zähne altersabhängige anatomische Veränderungen auftreten. Deren Kenntnisse sind bei erkrankten Pferden für den normal-anatomischen Vergleich nötig. Alle Längenmessungen der Pulpen und des Abstandes von Zahn und Infraorbitalkanal wurden dreimal hintereinander durchgeführt, um Messfehler zu minimieren. Die Pulpenmessungen anhand der protonengewichteten MRT-Bilder zeigten eine Größenabnahme in Korrelation zum Zahnalter. Zusätzlich veranschaulichten die MRT-Aufnahmen, dass eine gemeinsame Pulpenhöhle bei jungen Pferden vorlag, bei einem mittleren Zahnalter von 2,025 und maximalen Zahnalter von 3,75 Jahren. Fast alle ausgewählten Zähne zeigten einen direkten Kontakt zum Sinus maxillaris caudalis (34/36). Einige (6/36) der ausgewählten Oberkieferbackenzähne lagen direkt am Infraorbitalkanal. Der Abstand zwischen Zahn und Canalis infraorbitalis wurde dreifach von einem Untersucher mit Hilfe transversaler CT-Schnittbilder gemessen und zeigte eine positive Korrelation zum Zahnalter des Pferdes.

Zusätzlich zu den genannten, postmortalen Untersuchungen, erfolgte bei drei der neun untersuchten Pferde eine CT- und MRT-Untersuchung am lebenden Pferd in Allgemeinanästhesie, und post mortem am aufgetauten Kopf nach Tiefgefrieren. Im Anschluss wurden dentale, periodontale sowie angrenzende Weichteilgewebe bei jedem der drei Pferde jeweils in vivo, post mortem und am aufgetauten Präparat beurteilt und intraindividuell verglichen, um festzustellen, ob der Präparatzustand einen Einfluss auf die CT- oder MRT-Bildqualität hat. Hierbei ergaben sich für die MRT signifikant bessere Bildqualitäten für das periodontale Ligament und die Schleimhaut des Sinus maxillaris caudalis in den Sequenzen der lebenden Pferde gegenüber denen der euthanasierten und aufgetauten Pferde. Trotz dieser Qualitätsabnahme wiesen die bewerteten Strukturen bis auf die Sinusschleimhaut

auch im Zustand post mortem und den wieder aufgetauten Präparaten eine gute MRT-Darstellbarkeit auf.
Um bei zukünftigen Studien oder Patienten in vivo dasselbe Protokoll anwenden zu können, wurden die CT- und MRT-Einstellungen so gewählt, dass eine praxistaugliche Anästhesiezeit eingehalten und trotzdem eine zufriedenstellende Bildauflösung erzielt werden konnte.
Nach den Ergebnissen der vorliegenden Arbeit zeigen sich deutliche altersabhängige Veränderungen der Anatomie, die mit Hilfe der CT und MRT abgebildet werden können. Dabei zeigt die Auswertung der beurteilten Strukturen, dass sich die beiden Bildgebungsverfahren in ihren Eigenschaften und der Darstellbarkeit der Strukturen komplementieren. Mit Hilfe der aus dieser Studie gewonnenen Erkenntnisse über die normal-anatomische Darstellbarkeit dentogener Strukturen, sowie der altersabhängigen Veränderungen, können pathologische Prozesse besser von physiologischen und altersabhängigen Normzuständen abgegrenzt werden. Zusätzlich dienen die Ergebnisse als Hilfsmittel für die Wahl des bestmöglichen bildgebenden Verfahrens für die jeweiligen Strukturen, die untersucht werden sollen. Möglicherweise könnten auch intra- oder postoperative Komplikationen bei endodontischen Behandlungen zukünftig mit Hilfe einer MRT-Untersuchung reduziert werden, in dem vorab Kenntnisse über die Vitalität und Beteiligung des Zahnhalteapparates an einer Zahnerkrankung gewonnen werden. Hierfür sind jedoch weitere Untersuchungen zur Darstellung pathologischer Zahnbefunde im 3-Tesla-MRT nötig.

6 Summary

Christin Röttiger

Age-related comparison of computed tomography and high field (3.0 T) MRI of equine cheek teeth and neighboring structures: comparative study of image quality in horses *in vivo*, post-mortem and frozen-thawed

Modern imaging techniques such as CT and MRI provide the advantage to produce images without superimpositions. Whereas CT is a well-established technique for equine dental diagnostics, dental MRI examinations are rarely used. The direct comparison of both imaging techniques made in the current study, could serve as knowledge base for the depiction of equine cheek teeth and adjacent tissues.

To evaluate the image quality, the heads of nine warmblood horses were examined with CT and 3-T MRI within six hours after euthanasia. A total of 36 maxillary cheek teeth were examined. The sample pool included eighteen Triadan 08s and eighteen Triadan 09s. The horses were first placed on a stationary CT table in right lateral recumbency and afterwards in dorsal recumbency on a non-stationary MRI table. Dorsal and transversal scan series of the head were acquired. The MRI sequences assessed were: T1 weighted (3D), T2 weighted, proton density weighted and fat-suppressed images (dorsal and transversal).

After image acquisition, CT and MRI slices from different planes of the cheek teeth and adjacent structures were chosen for the evaluation and comparison of the CT and MR images. Predefined anatomical landmarks were used to ensure comparability of the selected slices. A modified four-point rating scale was used by each of three veterinarian observers to evaluate the image quality as well as the visibility of the selected maxillary cheek teeth and neighboring structures in a total of 1080 images.

Results of the current study show that CT was the best imaging technique to portray hard dental and bony tissues. MRI gave an excellent depiction of soft endo- and periodontal units. Comparing the different MR sequences, proton-density weighted scans showed the best visibility for the pulps and the periodontal ligament. Knowledge of age-related changes is essential for diagnosis, as cheek teeth and surrounding structures alter with increasing age.

Not only the depiction of the cheek teeth and their adjacent tissues, but also age-dependent variances were analyzed. Knowledge of age-related changes is essential for diagnosis, as cheek teeth and surrounding structures alter with increasing age. The positional relations between the dental roots and the floor of the paranasal sinuses were described (with CT and MRI), pulpar sizes (with MRI) and the distance between the dental alveoli and the infraorbital canal (with CT) were measured. All measurements were repeated three times. Negative correlation between dental age and pulpar dimensions was found. The median dental age for cheek teeth with a common pulp chamber was 2.025 years and the maximum age was 3.75 years. Most of the selected cheek teeth were in contact with the maxillary sinus (34/36). Some of them showed direct contact with the infraorbital canal (6/36). The distance between the infraorbital canal and the selected teeth increased with dental age.

The second objective of the current study was to quantify the impact of specimens' conditions (alive, postmortem, frozen-thawed) on the image quality in CT and MRI. Therefore, three horses of the study population were additionally examined *in vivo* in general anaesthesia and post-mortem after freezing and thawing. After image acquisition, CT and MRI slices from different planes of the cheek teeth and adjacent structures were chosen (Table 1). Three slices through each of the maxillary 08s and 09s were selected in a dorsal and transversal orientation in the CT, T2w, PDw and PD SPAIR scans. The comparisons of the image quality between live, post-mortem and frozen-thawed specimens showed that image quality parameters did not suffer post-mortem or by freezing and thawing; the image sharpness was even better in these groups than in live horses and visibility scores were satisfactory for soft tissues in all specimen conditions. However, significant superiority to portray the mucosa of the sinuses and the PDL was present in live horses.

The present study provides information about the dental and periodontal age-related morphology and its visibility via different imaging techniques in clinically healthy horses. Both 3T MRI and CT provide a valuable tool for the visualization and detection of dental and neighboring tissues. Both modalities complement each other, because MRI highlights soft dental and adjacent soft structures and CT depicts hard dental and bony surrounding tissues. The results aid in evaluating CT and MR images and in choosing the superior imaging modality. Although MRI is not applied as a routine diagnostic measure in dental pathologies, it provides some advantages that could be used for the detection of pulpar changes or before endodontic procedures. Future investigation is required to prove the opportunities of detecting and distinguishing pathological processes comparing MR and CT imaging. Knowledge about the age-related positional relations of the maxillary 08 s and 09 s and their adjacent structures expanded with imaging modalities (MRI, CT) might help to evaluate further clinical cases. False positive or negative results of dental pathologies can be avoided through optimal imaging. Additionally, intra- or postoperative complications can be prevented, as surgical planning is optimized.

7 Literaturverzeichnis

ARENCIBIA A., J. M. VAZQUEZ, R. JABER, F. GIL, J.A. RAMIREZ, M. RIVERO, N. GONZALEZ u. E. R. WISNER (2000):
Magnetic resonance imaging and cross-sectional anatomy of the normal equine sinuses and nasal passages.
Vet. Radiol. Ultrasound 41, 313-319

BAKER, G. J. (1999):
Dental decay and endodontic disease.
In: BAKER GJ, EASLEY J (editors): Equine dentistry.
1st ed., WB Saunders, London, S. 79-84

BARAKZAI, S. Z. (2011):
Dental imaging.
In: J. EASLEY, P. M. DIXON u. J. SCHUMACHER (Hrsg.): Equine Dentistry
3. Aufl., Saunders, Elsevier, Philadelphia, S. 199-230

BARBEE, D.D., J. R. ALLEN u. P.R. GAVIN (1987):
Computed tomography in horses.
Vet. Rad. 28(5), 144-151

BIENERT, A. (2002):
Digitalradiographische, computertomographische und mikrobiologische Untersuchungen bei Backenzahnerkrankungen des Pferdes.
Hannover, Stiftung Tierärztl. Hochschule, Klinik f. Pferde, Diss.

BOLAS, N. (2010):
Basic MRI principles.
In: R. C. MURRAY: Equine MRI 1,
Blackwell Publishing Ltd, Chichester, West Sussex, S. 3-37

BOLEN, G. E., D. HAYE, R. F. DONDELINGER, L. MASSART u. V. BUSONI (2011):
Impact of successive freezing-thawing cycles on 3-T magnetic resonance images of the digits of isolated equine limbs.
AJVR. 72(6), 780-790

BRINKSCHULTE, M. (2012):
Morphologische Untersuchung der Apertura nasomaxillaris des Pferdes sowie deren Verzweigung in die Nasennebenhöhlen unter Anwendung dreidimensionaler Rekonstruktion computertomographischer Schnittbildserien.
Hannover, Stiftung Tierärztl. Hochschule, Klinik f. Pferde, Diss.

BÜHLER M., A. FÜRST, F. I. LEWIS, M. KUMMER u. S. OHLERTH (2014):
Computed tomographic features of apical infection of equine maxillary cheek teeth: a retrospective study of 49 horses.
Equine Vet. J. 46, 468-73

CAVALLERI, J. M., J. METZGER, M. HELLIGE, V. LAMPE, K. STUCKENSCHNEIDER, A. TIPOLD, A. u. K. FEIGE (2013):
Morphometric magnetic resonance imaging and genetic testing in cerebellar abiotrophy in Arabian horses.
BMC Vet. Res. 9(1), 105

CASEY, M. B., G. R. PEARSON, J. D. PERKINS u. W. H. TREMAINE (2015):
Gross, computed tomographic and histological findings in mandibular cheek teeth extracted from horses with clinical signs of pulpitis due to apical infection.
Equine Vet. J. 47(5), 557-567

CHANG, K. J., I. R. KAMEL, K. J. MACURA u. D. A. BLUEMKE (2008):
3.0-T MR imaging of the abdomen: comparison with 1.5 T.
Radiographics 28(7), 1983-1998

DACRE, I., S. KEMPSON u. P. M. DIXON (2007):
Equine idiopathic cheek teeth fractures. Part 1: Pathological studies on 35 fractured cheek teeth.
Equine Vet. J. 39, 310-318

DACRE, I. T., S. KEMPSON u. P. M. DIXON (2008):
Pathological studies of cheek teeth apical infections in the horse. 1. Normal endodontic anatomy and dentinal structure of equine cheek teeth.
Vet J. 178, 311-20

DAKIN, S. G., R. LAM, E. REES, C. MUMBY, C. WEST u. R. WELLER (2014):
Technical set-up and radiation exposure for standing computed tomography of the equine head.
Equine Vet. Educ. 26(4), 208-215

DIXON, P. M. (1997):
Dental extraction in horses: Indications and praeoperative evaluation.
Compend. Contin. Educ. 119(3), 366-375

DIXON, P. M. (2011):
Dental anatomy.
In: J. EASLEY, P. M. DIXON u. J. SCHUMACHER (Hrsg.): Equine Dentistry
3. Aufl., Saunders, Elsevier, Philadelphia, S. 51-76

FLOYD, M. (1991):
The modified Triadan system: nomenclature for veterinary dentistry.
J. Vet. Dent. 8, 18-9

FREEMAN, S. L. u. G. C. ENGLAND (2000):
Investigation of romifidine and detomidine for the clinical sedation of horses.
Vet. Rec. 147(18), 507-511

GAUDINO, C., R. COSGAREA, S. HEILAND, R. CSERNUS, B. BEOMONTE ZOBEL, M. PHAM, T. S. KIM, M. BENDSZUS u. S. ROHDE (2011):
MR-imaging of teeth and periodontal apparatus: an experimental study comparing high-resolution MRI with MDCT and CBCT.
Eur Radiol. 21, 2575-83

GEIBEL, M., E. SCHREIBER, A. BRACHER, E. HELL, J. ULRICI, L. SAILER, Y. OZPEYNIRCI u. V. RASCHE (2015):
Assessment of apical periodontitis by MRI: a feasibility study.
RöFo. 187, 269-75

GERLACH, K., K. FLATZ, W. BREHM u. J. SEEGER (2009):
Klinische Anatomie des Gesichtsbereiches des Pferdes in der Magnetresonanztomographie.
Pferdeheilkunde 25, 45-52

GERLACH, K. u. GERHARDS H. (2008):
Magnetresonanztomographische Merkmale von Zubildungen im Bereich der Nase, Nasennebenhöhlen und der angrenzenden Knochen: retrospektive Analyse von 33 Pferden.
Pferdeheilkunde 24, 565-76

GERLACH, K., E. LUDEWIG, W. BREHM, H. GERHARDS u. U. DELLING (2013):
Magnetic resonance imaging of pulp in normal and diseased equine cheek teeth.
Vet. Radiol. Ultrasound 54(1), 48-53

HENNINGER, W., E. M. FRAME, M. WILLMANN, H. SIMHOFER, D. MALLECZEK u. S. M. KNEISSL (2003):
CT features of alveolitis and sinusitis in horses.
Vet. Radiol. Ultrasound 44(3), 269-76

HONTOIR, F., J. F. NISOLLE, H. MEURISSE, V. SIMON, M. TALLIER, R. VANDERSTRICHT, N. ANTOINE, J. PIRET, P. CLEGG u. J. M. VANDEWEERD (2013):
A comparison of 3-T magnetic resonance imaging and computed tomography arthrography to identify structural cartilage defects of the fetlock joint in the horse.
Vet J. 199, 115-122

ILLENBERGER, N., W. BREHM, E. LUDEWIG u. K. GERLACH (2013):
Darstellung altersabhängiger Veränderungen der Zahnpulpen ausgewählter Oberkieferbackenzähne des Pferdes mittels Magnetresonanztomographie.
Pferdeheilkunde. 29, 183-8

JOHNSTON, G. M., J. K. EASTMENT, J. L. N. WOOD u. P. M. TAYLOR (2002):
The confidential enquiry into perioperative equine fatalities (CEPEF): mortality results of phases 1 and 2.
Vet. Anaesth. Analg. 29,159-70

KAMINSKY, J., A. BIENERT-ZEIT, M. HELLIGE u. B. OHNESORGE (2014):
3 tesla magnetic resonance imaging of the equine nasal cavities, paranasal sinuses and adjacent anatomical structures in 13 healthy horses.
Pferdeheilkunde. 29, 183-8

KAMINSKY, J., A. BIENERT-ZEIT, M. HELLIGE, K. ROHN u. B. OHNESORGE (2016):
Comparison of image quality and in vivo appearance of the normal equine nasal cavities and paranasal sinuses in computed tomography and high field (3.0 T) magnetic resonance imaging.
BMC Vet. Res. 12(1), 13

KATTI, G., S. A. ARA u. A. SHIREEN (2011):
Magnetic resonance imaging (MRI)–A review.
Internat. J. Dent. Clin. 3, 65-70

KOPKE, S., N. ANGRISANI u. C. STASZYK (2012):
The dental cavities of equine cheek teeth: three-dimensional reconstructions based on high resolution micro-computed tomography.
BMC Vet. Res. 8, 1-16

KRAFT, S. L. u. P. GAVIN (2001):
Physical principles and technical considerations for equine computed tomography and magnetic resonance imaging.
Vet. Clin. North Am. Equine Pract. 17, 115-130

KRASNY, A., N. KRASNY u. A. PRESCHER (2012):
Anatomic variations of neural canal structures of the mandible observed by 3-tesla magnetic resonance imaging.
Journal of computer assisted tomography 36(1), 150-153

KRESS, B., A. GOTTSCHALK, L. ANDERS, C. STIPPICH, F. PALM, W. BÄHREN u. K. SARTOR (2004):
Highresolution dental magnetic resonance imaging of inferior alveolar nerve responses to the extraction of third molars.
Eur Radiol. 14, 1416-20

LANDIS, J. R. u. G. G. KOCH (1977):
An application of hierarchical kappa-type statistics in the assessment of majority agreement among multiple observers.
Biometrics, 363-374

LUKASIK, V. M., R. D. GLEED, J. M. SCARLETT, J. W. LUDDERS, P. F. MOON, J. L. BALLENSTEDT u. A. T. STURMER (1997):
Intranasal phenylephrine reduces post anaesthetic upper airway obstruction in horses.
Equine Vet. J. 29(3), 236-238

LUNDSTRÖM, T. S. (2012):
Orthograde endodontic treatment of equine teeth with periapical disease: long-term follow-up.
In: Proceedings of the 51st British Equine Veterinary Association. S. 105-106

LUNDSTRÖM, T. u. O. WATTLE (2016):
Description of a technique for orthograde endodontic treatment of equine cheek teeth with apical infections.
Equine Vet. Educ. 28, 1-12

MAYRHOFER, E. u. W. HENNINGER (1995):
Computertomographie in der Veterinärmedizin.
Veterinärspiegel 1, 14-22

MERKLE, E. M. u. B. M. DALE (2006):
Abdominal MRI at 3.0 T: the basics revisited.
American Journal of Roentgenol, 186(6), 1524-1532

MORROW, K. L., R. D. PARK, T. L. SPURGEON, T. S. STASHAK u. B. ARCENEAUX (2000):
Computed tomographic imaging of the equine head.
Vet. Radiol. Ultrasound 41(6), 491-497

NICKEL, R., A. SCHUMMER, K. H. WILLE u. K. WILKENS (2004):
Knochenlehre, Osteologia.
In: R. NICKEL, A. SCHUMMER und E. SEIFERLE Lehrbuch der Anatomie der Haustiere
8. Aufl., Parey, Berlin, S. 15-214

PARENTE, E. J., S. H. FRANKLIN, F. J. DERKSEN, M. A. WEISHAUPT, H. J. CHALMERS u. C. TESSIER (2011):
Diagnostic techniques in equine upper respiratory tract disease.
In: J. A. AUER und J. A. STICK: Equine Surgery
4th ed., Elsevier, Saunders, St. Louis, Missouri, S. 536-556

ROS, K. (2011):
Röntgentechnik und -befunde.
In: C. VOGT (Hrsg.): Lehrbuch der Zahnheilkunde beim Pferd
Schattauer, Stuttgart, S. 65-98

SAUPE, N., K. P. PRÜSSMANN, R. LUECHINGER, P. BÖSIGER, B. MARINCEK u. D. WEISHAUPT (2005):
MR imaging of the wrist: comparison between 1.5- and 3-T MR imaging - preliminary experience.
Radiology 234(1), 256-264

SCHWARZ, T. u. J. SAUNDERS (2011):
Veterinary computed tomography.
1st ed., John Wiley & Sons Ltd., West Sussex

SIMHOFER, H., C. STOIAN u. K. ZETNER (2008):
A long-term study of apicoectomy and endodontic treatment of apically infected cheek teeth in 12 horses.
The Vet. J. 178(3), 411-418

STASZYK, C. (2015):
Zähne und Gebiss des Pferdes – eine klinisch-anatomische Übersicht.
Tierärztliche Praxis G: Großtiere/Nutztiere 43(06), 375-386

STOLL, M. u. C. PEARCE (2017):
Extraoral endodontal treatment and replantation of equine cheek teeth. Part 2: Technique and materials.
Proceedings of the 26th European Veterinary Dental Forum 2017
Malaga, S. 121-123

STRAUCH, S., J. SCHWARZER u. S. KAU (2019):
Computertomografie in der Pferdezahnheilkunde – ein Überblick.
Pferdespiegel 22(1), 23-36

TREMAINE, W. H. u. P. M. DIXON (2001):
A long-term study of 277 cases of equine sinonasal disease. Part 1: details of horses, historical, clinical and ancillary diagnostic findings.
Equine Vet. J. 33, 274-82

TUCKER, R.L. u. S. N. SAMPSON (2007):
Magnetic resonance imaging protocols for the horse.
Clin. Tech. Equine Pract. 6, 2-15

TYMOFIYEVA, O., P. C. PROFF, K. ROTTNER, M. DÜRING, P. M. JAKOB u. E. J. RICHTER (2013):
Diagnosis of dental abnormalities in children using 3-dimensional magnetic resonance imaging.
J. Oral Maxillofac. Surg. 71, 1159-1169

VALLANCE, S. A., R. J. BELL, M. SPRIET, P. H. KASS u. S. M. PUCHALSKI (2011):
Comparisons of computed tomography, contrast enhanced computed tomography and standing low-field magnetic resonance imaging in horses with lameness localized to the foot. Part 1: anatomic visualization scores.
Equine Vet. J. 44, 51-56

VANDENBERGHE, B., R. JACOBS u. H. BOSMANS (2010):
Modern dental imaging: a review of the current technology and clinical applications in dental practice.
Eur. Radiol. 20, 2637-2655

VERAA, S., G. VOORHOUT u. W. R. KLEIN (2009):
Computed tomography of the upper cheek teeth in 422 horses with infundibular changes and apical infection.
Equine Vet. J. 41, 872-6

WAGNER, A. E. (2008):
Complications in equine anesthesia.
Vet Clinics of North America: Equine Practice. 24, 3, 735-52

WERPY, N. M. (2007):
Magnetic Resonance Imaging of the Equine Patient: A Comparison of High- and Low-Field Systems.
Clinical Techniques in Equine Practice. 6, 37-45

WINDLEY, Z., R. WELLER, W. H. TREMAINE u. J. D. PERKINS. (2009):
Two- and three-dimensional computed tomographic anatomy of the enamel, infundibulae and pulp of 126 equine cheek teeth. Part 2: findings in teeth with macroscopic occlusal or computed tomographic lesions.
Equine Vet. J. 41, 441-7

8 Anhang

Anhang 1 (Appendix 1, Manuskript II): Imaging techniques and parameters

Imaging technique	Sequence	Orientation	Mean scan duration in minutes	Matrix	TR (ms)	TE (ms)	ST (mm)
MRI	T1w	dorsal (MPR)	14	1024 x 1024	8.5	3.9	0.9
	T2w	transverse	18	1024 x 1024	4500	80	3.1
	T2w	dorsal	10	1024 x 1024	3000	80	4
	PDw	dorsal	16	1024 x 1024	6400	30	4
	PD SPAIR	transverse	24	960 x 960	8872	30	3
CT	Axial Scan	transverse (MPR)	3	1024 x 1024			2

MPR: Multiplanar reconstruction, TE: echo time, TR: time to repetition, ST: slice thickness

9 Erklärung über die erbrachten Leistungen

Gemäß § 8 Absatz 3 der Promotionsordnung der Tierärztlichen Hochschule Hannover hat der Promovent bei einer Dissertation, die auf Veröffentlichungen basiert (kumulative Dissertation), an denen mehrere Autoren beteiligt waren, den selbstständigen Anteil an den vorgelegten Publikationen darzulegen.

- Auswahl der MRT-Sequenzen (in Zusammenarbeit mit Dr. Maren Hellige)
- Durchführung der CT- und MRT-Untersuchungen (in Zusammenarbeit mit Dr. Maren Hellige)
- Betreuung der Pferde vor und nach den Untersuchungen (in Zusammenarbeit mit den Mitarbeitern der Klinik für Pferde der Stiftung Tierärztlichen Hochschule Hannover
- Aufbereitung des akquirierten Bildmaterials
- Auswertung des angefertigten Bildmaterials (in Zusammenarbeit mit Dr. Karl Rohn)
- Beschreibung der dargestellten anatomischen Strukturen
- Messungen der Pulpengröße, des Abstandes zwischen Zahn und Canalis infraorbitalis
- Auswahl des Bildmaterials zur Auswertung durch Gutachter
- Ausarbeitung des Auswertungsschlüssels
- Begutachtung des Bildmaterials als einer der drei Gutachter
- Aufbereitung der Ergebnisse der Bildbegutachtung
- Auswahl der Publikationsthemen
- Anfertigen der Publikation/ der Manuskripte

10 Danksagung

Im Folgenden danke ich den Personen, die mich bei der Anfertigung dieser Arbeit durch besonderes Engagement unterstützt haben.

Mein Dank gilt an dieser Stelle Prof. Dr. Bernhard Ohnesorge für die stets freundliche und konstruktive Unterstützung während der gesamten Doktorarbeit.

Ganz herzlich möchte ich mich auch bei meiner wissenschaftlichen Betreuerin Frau PD Dr. Astrid Bienert-Zeit für die fachliche Unterstützung, die Auswertung der zahlreichen Bilder, die konstruktiven Korrekturen und ihre Geduld bedanken.

Ein besonderer Dank gilt Frau Dr. Maren Hellige für ihre fachliche und konstruktive Unterstützung in Zusammenhang mit der Doktorarbeit, ihre Hilfe, Einarbeitung und Zeit bei den computertomographischen und magnetresonanztomographischen Untersuchungen sowie für ihre zeitaufwendigen Auswertungen der CT- und MRT-Bilder.

Bedanken möchte ich mich auch bei Frau Dr. Elisabeth Engelke für die fachliche Unterstützung und bei Herrn Oliver Stünkel für die Bereitstellung von Kopfpräparaten für Vorversuche.

Bei allen Mitarbeitern der Klinik für Pferde bedanke ich mich für die große Hilfsbereitschaft. Vor allem für eure tatkräftige Unterstützung bei den CT- und MRT-Aufnahmen, sowie die stets motivierenden Worte bin ich sehr dankbar.

Von ganzem Herzen möchte ich mich bei meiner Familie und meinen Freunden für ihre moralische Unterstützung bedanken. Um meine Doktorarbeit fertigstellen zu können, war es ein wichtiger Grundstein für mich, zu wissen, dass wir selbst in den schwersten Zeiten zusammenhalten. Großer Dank gilt dabei auch besonders meinen

Eltern und Großeltern für die finanzielle Hilfeleistung während der gesamten Ausbildung.

Mein größter Dank gilt meinem Ehemann für die Hilfe bei Computerproblemen, die endlose Geduld, die positive Art und das Verständnis während der gesamten Promotionszeit. Danke Lars, für dich und unser Leben mit unserer kleinen Tochter Liv Sophie.